The Johnstone Triangle
The Key to Understanding Chemistry

**Advances in Chemistry Education Series**

*Editor-in-chief:*
Keith S. Taber, *University of Cambridge, UK*

*Series editors:*
Avi Hofstein, *Weizmann Institute of Science, Israel*
Vicente Talanquer, *University of Arizona, USA*
David Treagust, *Curtin University, Australia*



*Titles in the Series:*
1: Professional Development of Chemistry Teachers: Theory and Practice
2: Argumentation in Chemistry Education: Research, Policy and Practice
3: The Nature of the Chemical Concept: Re-constructing Chemical Knowledge in Teaching and Learning
4: Creative Chemists: Strategies for Teaching and Learning
5: Engaging Learners with Chemistry: Projects to Stimulate Interest and Participation
6: The Johnstone Triangle: The Key to Understanding Chemistry

*How to obtain future titles on publication:*
A standing order plan is available for this series. A standing order will bring delivery of each new volume immediately on publication.

*For further information please contact:*
Book Sales Department, Royal Society of Chemistry, Thomas Graham House, Science Park, Milton Road, Cambridge, CB4 0WF, UK
Telephone: +44 (0)1223 420066, Fax: +44 (0)1223 420247,
Email: booksales@rsc.org
Visit our website at books.rsc.org

# *The Johnstone Triangle*
## *The Key to Understanding Chemistry*

By

**Norman Reid**
*University of Glasgow, UK*
*Email: dr_n@btinternet.com*

ROYAL SOCIETY
OF **CHEMISTRY**

Advances in Chemistry Education Series No. 6

Print ISBN: 978-1-83916-168-1
PDF ISBN: 978-1-83916-366-1
EPUB ISBN: 978-1-83916-367-8
Print ISSN: 2056-9335
Electronic ISSN: 2056-9343

A catalogue record for this book is available from the British Library



The Royal Society of Chemistry is a charity, registered in England and Wales, Number 207890, and a company incorporated in England by Royal Charter (Registered No. RC000524), registered office: Burlington House, Piccadilly, London W1J 0BA, UK, Telephone: + 44 (0) 20 7437 8656.

For further information see our website at www.rsc.org

For general enquiries, please contact books@rsc.org

For EU product safety enquiries, please email books@rsc.org or contact Royal Society of Chemistry Worldwide (Germany) GmbH, Römischer Hof, Unter den Linden 10, 10117 Berlin.

# *Preface*

This book focusses on the triangle model that Alex H. Johnstone developed in the early 1980s and which has appeared in many publications since that time. He originally conceived the model in the context of making chemistry more accessible to a wider range of learners, but it has been applied in almost every area of chemistry education at all stages of learning.

In his sustained research, where he was exploring why understanding chemistry is demanding, Johnstone asked two question. First, does the problem relate to the nature of chemistry and, second, does it relate to the way humans gain understanding? Both questions were found to be important and the answers were found to be connected.

The connection involves the research about working memory, the place in the brain where we think, understand and solve problems. Its fixed mature and limited capacity are controlling factors on all understanding. Understanding chemistry is cognitively demanding in that it places heavy burdens on limited cognitive resources.

The first three chapters tell the story of how the triangle model came to exist and the way the model rationalises so much evidence from chemistry education research as well as the repeated experiences of teachers of chemistry at all levels. The central finding from research is the critical role of working memory.

The next three chapters look at the implications of the model for the way a chemistry curriculum should be constructed and the processes of teaching and learning in chemistry when the goal is focussed on understanding the key ideas. With an emphasis on understanding, it is important that learners develop valid *mental* models that can interpret experimental observations and make sense of the world around them at the molecular level and then communicate these models in accessible ways.

---

Advances in Chemistry Education Series No. 6
The Johnstone Triangle: The Key to Understanding Chemistry
By Norman Reid

Published by the Royal Society of Chemistry, www.rsc.org

At that point, one chapter looks at examples of the many practical tips and tricks that can help learners in this task, many of them drawn from the work of Johnstone himself. In all of this, the value of the triangle model is illustrated, showing how it can point to better ways forward in the teaching and learning of chemistry. The final chapter brings together all the evidence from research to present some practical pointers for teaching.

This book is a research story. It is *not* based on anecdotal or idiosyncratic experience. It seeks to consider the clear evidence, confirmed in numerous research studies throughout the world, that reveals the critical importance of the functioning of working memory in achieving sound understanding. The triangle model is a brilliantly simple way to summarise key ideas as well as offering clear directions for all involved in chemistry education. It is hoped that this will prove helpful to all involved in sharing the exciting story of the way humans have come to understand the molecular world, one of the great examples of incredible human endeavour.

Norman Reid

# *Acknowledgements*

I would like to thank the Royal Society of Chemistry (RSC), and Professor Keith Taber specifically, for inviting me to write this text and for all the help and encouragement given. I would also like to thank Michelle Carey and Lewis Pearce from the RSC who have guided me though the processes involved with patience, a characteristic I often lack!

Over the years, many students have encouraged me as they developed a love of learning, as they applied their able minds to difficult ideas and as they challenged me to think more and more about the complex nature of learning. School students have never ceased to amaze me in what they can achieve, given just a little encouragement. University research students have undertaken demanding research studies with a level of competence and insight that has challenged. Their findings have played a major role in the development of this book and I am grateful to them for their commitment, their sheer intellectual power, and the fun and laughter we have enjoyed together as we travelled a road of educational enquiry together.

Norman Reid

Advances in Chemistry Education Series No. 6
The Johnstone Triangle: The Key to Understanding Chemistry
By Norman Reid

Published by the Royal Society of Chemistry, www.rsc.org

# *Dedication*

*Dedicated to the memory of*
*the late Professor Alex H. Johnstone*
*who inspired so many over so many decades*
*in their love of chemistry.*

Advances in Chemistry Education Series No. 6
The Johnstone Triangle: The Key to Understanding Chemistry
By Norman Reid

Published by the Royal Society of Chemistry, www.rsc.org

# *The Author*

**Norman Reid, Emeritus Professor, University of Glasgow, Scotland**
Norman has taught chemistry at both secondary and university levels as well as having been a headteacher. He then became director of a major international research centre in science education, involving the supervision of large numbers of research students. He held chairs in two Scottish universities. A frequent international keynote speaker, he has published very extensively. Areas of specialist interest include educational research, science education, attitude development, assessment, learner characteristics, educational statistics and quality assurance.

Advances in Chemistry Education Series No. 6
The Johnstone Triangle: The Key to Understanding Chemistry
By Norman Reid

Published by the Royal Society of Chemistry, www.rsc.org

# Contents




Advances in Chemistry Education Series No. 6
The Johnstone Triangle: The Key to Understanding Chemistry
By Norman Reid



Published by the Royal Society of Chemistry, www.rsc.org

CHAPTER 1

# *The Working Memory Story: Why Is It Difficult to Understand Chemistry?*

## 1.1 The Early Days

Alex H. Johnstone was born in the port town of Leith, which is now absorbed as part of the city of Edinburgh. He was a member of a large family and attended the local secondary school where he did extremely well, gaining entrance to the University of Edinburgh. Originally planning to take a degree in Physics, he graduated with a First-Class Honours degree in Chemistry, his greatest interest being in analytical chemistry.[†]

After a year in teacher training and a period of mandatory army service, he entered secondary school teaching in a school in Edinburgh. Here, he taught chemistry from ages 12 to 18 for a number of years before being appointed to the post of Principal Teacher of Chemistry[‡] in a school in Stirling, central Scotland. During the period of his school teaching, major changes were taking place in the teaching of chemistry (as well as physics and mathematics) in many countries.

Alex was driven by a desire to be the best he could be, no matter what the challenges of life threw at him. He was also driven by an intense questioning curiosity. He wanted to know. He wanted to understand. He wanted to make sense of all that he had studied in chemistry. He wanted to share that

[†]Like most European countries, Scottish degrees last four years and allow enormous flexibility, permitting changes in main degree subjects to be made *during* a degree.

[‡]The teaching profession is organised very differently in Scotland when compared to England. It has always been an all-graduate profession and a Principal Teacher of Chemistry at that time had responsibility for a team of graduate chemists.

Advances in Chemistry Education Series No. 6
The Johnstone Triangle: The Key to Understanding Chemistry
By Norman Reid

Published by the Royal Society of Chemistry, www.rsc.org

understanding with learners. This restless enquiring mind was highly creative and never left him throughout his life. However, there was another side to him. He was a natural teacher, able to explain and interpret the most complex ideas making them accessible to young learners. Alongside this, he really cared about his students. He was always ready to give them the time and support when they faced difficulties in understanding. The teaching and the support were offered with a deep sense of fun and humour, setting his students at ease and making them feel valued.

In the early years of his teaching in Edinburgh, he was searching in an old storeroom in the school. He came across a student notebook dating from the early 20th century. Blowing off the dust, he was horrified to find that what he was expected to teach in the late 1950s was more or less the same as what had been taught about 50 years before. This started him thinking. Chemistry research had made enormous progress over half a century. Despite this, little change was apparent in what was to be taught at school. To him, this undermined the real nature of chemistry, which was moving at the frontiers of knowledge and understanding in seeking to make sense of the world around.

With his driving enquiring mind, he set about looking for any books or other publications that might help. This was in the days before the development of the modern journals covering chemistry education and wider science education. However, there were some journals that offered collations of laboratory experiments for schools and all kinds of practical ideas that might enrich what was to be taught. He collated many of these and started to try them out. His creative mind started to develop more.

Somewhat speculatively, this led him to develop ideas for a radically new curriculum for chemistry in Scottish schools, leading to the Scottish Higher Grade examination.[§] However, he had no outlet to make his ideas known. Unexpectedly, an event in the United States started changes that provided him with that outlet. In October 1957, the Russians had launched their first satellite (sputnik) beating the Americans in the space race. The Americans were mortified and started to blame the inadequacies of their school maths-science education for this failure. In fact, there was little evidence for this assumption. It is typical that schools and teachers are often blamed for societal ills over which they have no control or influence!

This set in motion a series of curriculum upheavals in the United States, the aim being to update what was to be taught and to focus on skills related to understanding instead of recall. Early on in this discussion, an international conference was held to exchange ideas and was attended by a small group from Scotland. Scotland was ready for change. The school science base was strong but what was strong was very out-of-date.

During an inspection visit to the school in Edinburgh where Johnstone taught, the chief inspector for schools for chemistry shared with the three

---

[§]Since 1874, Scotland has had an examination system based on the Higher Grade, not dissimilar to many modern baccalaureate systems throughout the world. Typically, able students might sit six subjects and these were the basis for entry to Higher Education.

chemistry teachers in the school that some radical curriculum revisions were being considered. Johnstone, as a very young teacher, was present and, somewhat diffidently, offered to give his ideas. The inspector left with a bundle of papers outlining the new curriculum that Johnstone had developed.

After a short time, the inspector made a surprise return to the school. He asked if he had permission to use the ideas Johnstone had developed. In fact, he went further. He stated that he was going to '*throw away*' other ideas and develop the new curriculum based mainly on what Johnstone had submitted. In due time, the new Higher Grade Scottish Curriculum in Chemistry (and also Higher Grade Physics) was published in 1962, covering the entire secondary school age range up to age 17 (when the Higher Grade was sat). The first examination was offered in 1965 (Curriculum Papers 512, 1962).

Later in his career, Johnstone (1993) summarised the historical development of the teaching of chemistry. At Glasgow University, chemistry was first taught in the early 1750s, but the significance of chemistry education grew rapidly with the appointment of Joseph Black in 1756. Chemistry findings were beginning to make an impact on agriculture and industry and there was growing public interest in chemistry. Black made major contributions in all these areas during his time at Glasgow University and later at Edinburgh University. However, the training of chemists in universities focussed mainly on the development of personnel to support the developments in industry.

**Joseph Black (1728–1799)**
Professor of Chemistry, University of Glasgow (1756–1766).
Professor of Chemistry, University of Edinburgh (1766–1799).
Practising as a medical doctor, he made breakthroughs in relation to the nature of carbon dioxide. In 1756, Black met up with James Watt (steam engine fame). This stimulated his interest in the concept of latent heat and specific heats.
Black was an excellent lecturer, illustrating his lectures with elegant experimental demonstrations.
Former students moved to Chairs of Chemistry in Edinburgh, Glasgow, Cambridge and Oxford.

The teaching of chemistry rarely featured in schools until the latter end of the 19th century where '*it was allowed grudgingly for the lower classes who would earn their living with their hands*' (Johnstone, 1993, p. 702). The curriculum was built around '*preparations and properties*', along with a catalogue of definitions to be memorised. Laboratory work was mainly observing a few experiments along with a considerable amount of analytical work. Much energy was often expended pointlessly in generating laboratory reports. The chemistry taught up to the middle of the 20th century still reflected many of these features.

Historically, Scotland has always been very strong in the sciences (and related subjects like medicine and engineering). Despite the curricula in

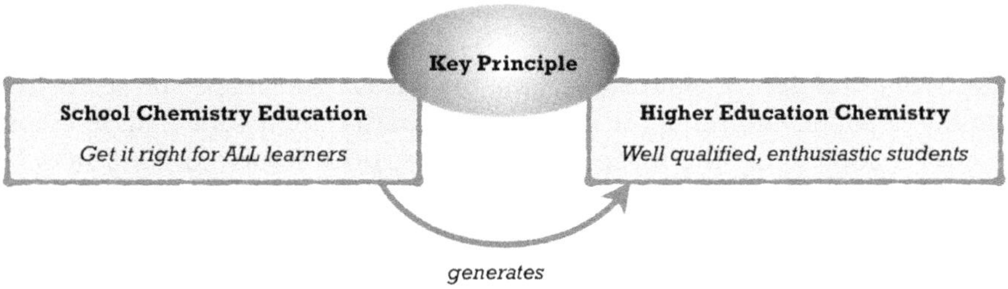


**Figure 1.1** The key principle.

chemistry and physics being demanding, the radical changes introduced in the revised Higher curricula contributed in making the sciences both increasingly popular at school level and, in due time, universities saw large numbers applying, a pattern that still is to be seen today. Indeed, the three sciences are usually the most popular elective subjects in the upper secondary school curriculum in Scottish schools. In the lower secondary school curriculum, the chemistry taught is regarded as being relevant and important for all learners.

The longer-term outcome is that, as a percentage of GDP, Scotland ranks very high for per capita scientific research spending.[¶] In fact, this lays down a key principle: getting the school science education right for all learners will generate well qualified and enthusiastic students for higher education, generating research leadership as needed (Figure 1.1). The later seminal work of Skryabina in physics revealed the key principles. We shall consider her findings later for they have major implications for curriculum design as well as the selection and training of chemistry teachers (Reid and Skryabina, 2002a, 2002b).

## 1.2 The 1960s Scottish School Chemistry Curriculum

The new curriculum was based on guided discovery and the central place of the laboratory was established. Indeed, chemistry was taught in laboratories in almost all schools throughout Scotland. The experiments were designed to test ideas, raise questions, explore new thinking. They were never employed to '*confirm theory*' and never seen as places to develop specific experimental skills. The goal was to explore the insights that chemistry might bring to enable learners to make sense of their world.

The approach of the new Scottish chemistry curriculum was radically different. Johnstone, along with his colleague Ian Morrison, were invited to train teachers. They spent considerable time offering training sessions for chemistry teachers throughout Scotland. However, the workload on both was very demanding. Both had full-time jobs teaching chemistry in the school in Stirling. The two then collaborated to write a series of textbooks (Johnstone and Morrison, 1964–1969) and this reduced the need for training meetings.

[¶]Although data are difficult to quantify, higher than most countries.

The chemistry curriculum has been revised several times since the 1960s and, sadly, some of the great features introduced by Johnstone have now been lost. They have usually been replaced with more traditional approaches. By contrast, a revision of the Physics curriculum in the late 1970s developed the exciting features even further and, for many years, physics was an exceptionally popular subject, with the retention rate[‖] being the *highest* for any elective subject in the Scottish curriculum. This is an amazing observation, given the low popularity for physics in many countries. A later study showed the very high standard that students were achieving in physics (Scottish Qualifications Authority, 1998). This illustrates another key principle: level of demand and popularity need not be related.

In the late 1960s, in his work on national committees, Johnstone met Professor David Sharp (1931–2013), the Ramsay Professor of Chemistry and Head of Department at Glasgow University. Professor Sharp was very committed to the enhancement of chemistry education at all levels and was looking for an able experienced teacher to head up a rigorous research programme. Johnstone was invited to move to Glasgow to undertake, and then lead, a sustained programme of fundamental research into the teaching and learning of chemistry and related subjects at both school and university levels. In due time, he was appointed to a professorial chair in the chemistry department. He was required to teach undergraduate chemistry, but his research all related to science education. This led to the formation of the Centre of Science Education which, in due time, moved into the Faculty of Science and widened its subject scope.

In the Centre, he set standards for research that matched those in the mainstream science disciplines. There was rigour, quantitative measurement, replication. He allocated the research projects to research students and, together, they formed a coherent programme of research that steadily revealed some of the fundamental principles that allowed learners to gain much more from their studies in chemistry and related subject areas. The work generated an apparently endless flow of publications, some of which are referenced in this book.

The work was characterised by his questioning and highly creative mind. Dialogues and debates with his research students have left indelible memories with them. His work had a great impact on university teaching and learning. Sadly, his work was largely ignored by education faculties and schools. Unfortunately, schoolteachers are rarely free to make changes in line with what his research had found but have to follow what is mandated for them. By contrast, university teachers have a greater measure of freedom to bring in new developments. The impact of his research has been enormous across university chemistry departments in many countries across the world. This can be seen in developments reported in many journals

[‖]The proportion who choose to move for one level to the next: in this case, from Standard Grade to Higher Grade—data gathered in the late 1990s.

(for example, a search through the highly cited journal: *Chemistry Education Research and Practice* reveals something of the extent of the impact).

Several years after his retirement, the Centre was flourishing, with up to 20 research students at times (mainly PhD and post-doc, with some research Masters) and an enormous publications output. Scottish sciences continued to flourish in Scottish schools. This was reflected in the considerable growth in the Faculty of Science at Glasgow. Re-organisation led the Faculty of Science being split into three more manageable Faculties. Around the same time, a new Faculty of Education was being established and the Centre for Science Education was then moved (somewhat against its wishes) into this Faculty of Education. Sadly, the Centre was closed by that Faculty on the retirement of Johnstone's successor, despite its success in attracting high quality postgraduate students from around the world and its sustained publications record.

## 1.3 Chemistry Is Difficult

Today, it seems to be well established across countries and cultures that subjects like chemistry and physics are really demanding for learners. Johnstone started his early work exploring this. The new Scottish chemistry curriculum was well established by the mid-late 1960s. However, reports started to come in that, while the learners enjoyed the course, they found parts of it demanding. His first task was to establish where the difficulties were. The findings were published in 1971 (Johnstone *et al.*, 1971). The key areas of difficulty were found to lie in areas like:

- Redox
- The mole
- Thermodynamics
- Organic chemistry (some areas)
- Areas where calculations were needed.

Later studies showed several difficult areas in physics, as well as one or two topics in biology.

He then set a series of research students to look at specific areas to try to find out *why* these topics were so demanding and what, if anything, might be done to improve the situation. Numerous useful findings were found but it was still not apparent *why* these areas posed the problems. However, in the late 1970s, a very able student (Natalie Kellett) was looking at organic chemistry and she had gathered and collated vast amounts of data from school learners. She laid out her findings and started to look for any patterns.

Suddenly, she started to see patterns that looked important. She observed that difficulties tended to occur in areas where the amount of information that the learners had to consider in *seeking understanding* was high. Was there some kind of *information overload*? She rushed out of her office and went looking for Johnstone, her supervisor. Immediately, he came back with her to look at what she was seeing. They spent some time looking at all the information. This insight seemed to offer a key.

She was nearly finishing her PhD and she completed her write-up and presented the observation in her thesis as a hypothesis (Kellett, 1978). Soon after, this was published as a paper (Johnstone and Kellett, 1980). The next PhD student (Hamdy El-Banna) was given the task of testing the hypothesis. It turned out that her hypothesis was supported, but in a rather surprising way. Johnstone had often wondered if the problems in understanding chemistry were related to the nature of chemistry as a subject or the nature of learning itself. Here was evidence that the nature of learning was one critical factor. It took a little longer to establish that the nature of chemistry was also a major problem and that the two were related.

## 1.4 Information Load

At the start, Kellett and Johnstone did not really appreciate the full meaning of the phrase '*information load*'. In a topic like organic chemistry, the sheer number of compounds and the bewildering variety of structures is very apparent, but this was NOT what was meant by information load. Slowly, it began to be understood that '*information load*' had to be seen in terms of the number of pieces of information that the learner has to hold in the mind *at the same time* to perform the task successfully or to gain understanding.

Information load is sometimes referred to as *cognitive load* and is nothing to do with the total amount of content to be covered. It relates to the total number of items of information, ideas and processes that are involved *at the same time* to be able understand something or to solve a problem. This is illustrated in Figure 1.2.

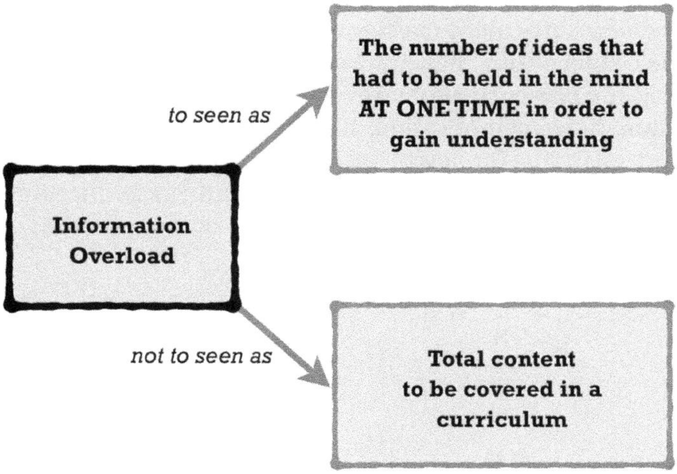


**Figure 1.2** Information overload.

We can see this in terms of an apparently simple reaction like esterification. Let us look at the synthesis of ethyl ethanoate:

$$C_2H_5OH + CH_3COOH \rightarrow CH_3COOC_2H_5 + H_2O$$

For novice learners, about the only thing they recognised with confidence was the formula for water. The equation had no obvious connections with what they undertaken in the laboratory. They had mixed very small quantities of ethanol and ethanoic acid and add a drop of concentrated sulfuric acid. Gently warming released the characteristic odour of the ester.

To make matters worse, the way it is written means that the ethanol *'bit'* in the product is written in reverse order in the product compared to the reactant. Indeed, novice learners may not yet be confident to understand what the COOH group meant. They are trying to remember what these various symbolisms mean, trying to relate the symbolisms to what they had undertaken in a test-tube in the laboratory and then they start to panic: how were they to remember all this for the examination?

One step forward might be show the reaction, employing what was called *'lasso chemistry'* (Figure 1.3).

School students were familiar with neutralisation and, sometimes, another approach was put forward. In this, it was suggested that the reaction was similar—the formation of water. Unfortunately, the alcohol acts as a proton donor and the acid generates the hydroxyl group for the water formation. The reaction in the test-tube requires a drop of concentrated sulfuric acid. Overall, the reaction is totally *unlike any neutralisation.* This approach is utterly misleading.

The sheer number of *'bits'* of information that the student has to hold in their minds *at the same time* to make sense of what is happening is large. It was little wonder that students were found to be memorising the equation in order to pass examinations. Understanding was the casualty.

What the new curriculum in chemistry had done, quite inadvertently, was to introduce topics with high information load earlier than before. More importantly, the new curriculum was seeking to generate understanding and to move away from memorisation. It seemed that some topics were simply beyond understanding for novice learners at that age. The question was why?

For the *'novice'* learner, the equation representing esterification may seem like a set of strange jumbled symbols that does not seem to relate to what they

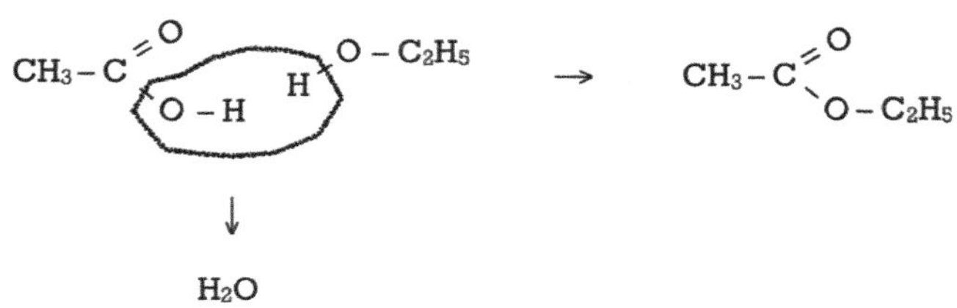


**Figure 1.3** Esterification: 'lasso' chemistry.

undertook in the laboratory. It all makes little sense to them. For the '*expert*', the equation carries great meaning. It is a useful summary of the chemical interactions involved. It can be seen as an interpretation. It can be seen in terms of resonance structures, equilibrium and three dimensions. By contrast, the '*novice*' learner is simply overwhelmed by information of limited meaning.

The great insight of Natalie Kellett suggested a pattern. Difficulty in some way was related to '*information load*' but exactly what that meant was not yet clear. Johnstone started to comb the literature to see if there was anything that gave any evidence about how the human brain handled information, particularly related to how much information the brain could handle *at the same time*. Some work had been published and this seemed to offer a key.

George Miller (1920–2012), a distinguished psychologist, had joined Jerome Bruner (1915–2016) to found the Centre for Cognitive Studies at Harvard University. The work of this Centre finally demolished the dominance of behaviourism and this placed cognition at the centre of research.

Before this, Miller had developed robust ways to measure the capacity of what he called the '*short-term memory*'. He published his findings in 1956 in a high-status journal, the paper carrying the most amazing title: '*The Magical Number Seven, Plus or Minus Two*' (Miller, 1956a). In simple terms, he showed how it was straightforward to measure the capacity of that part of the brain, which had been called the '*short-term memory*'. In any group of adults, the average capacity was found to be seven: the '*average*' individual could hold seven items of information simultaneously. However, we vary, and he found that most adults had short term memories with capacities ranging from five to nine.

The brilliance of Johnstone was that he saw how this finding offered some kind of rationale for what Kellett was seeing in her study in the understanding of organic chemistry. The learners could hold limited number of ideas *at the same time* in their short-term memories. Many ideas in organic chemistry require the learner to hold too many ideas and, as a result, understanding becomes impossible, unless they have a strategy to reduce the number of ideas.

Miller (1956a, 1956b) had also found another important feature. He realised that different people had different skills in the way they organised the information. Some could group items of information together and then the group could be seen as one. This reduces the pressure on the capacity of the short-term memory. The '*expert*' had developed the skills to do this, largely thorough experience. Therefore, the teacher found that the equation for esterification held enormous meaning. As a result of greater experience, the structures were seen as one, leaving space in the short-term memory for the meaning of chemical process to be considered. The novice learner lacked this experience, could not group items of information together meaningfully, and the short-term memory simply could not cope.

Miller described this using the word '*chunking*'. Chunking is the process of grouping ideas together so that they are seen as one. Miller referred to items of information as '*chunks*'. The question then was: was it the limited capacity of the short-term memory that was being overloaded with the amount of

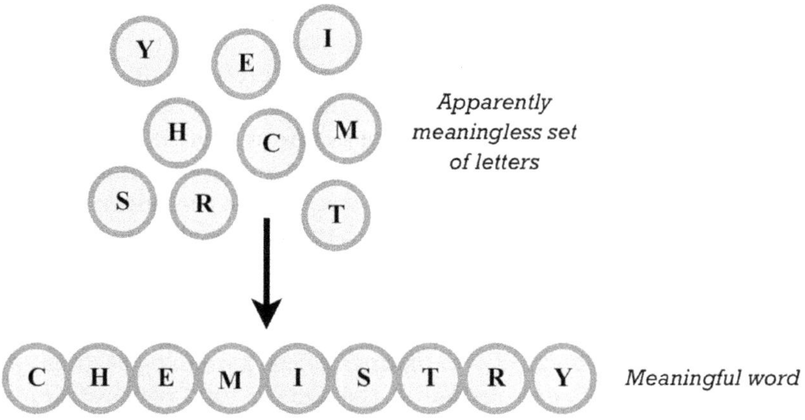


**Figure 1.4** Humans are pattern seekers.

information in topics that were proving difficult and that this was the source of the problem? Further research followed this up.

What Miller had found in his studies was that a '*chunk*' was what an individual saw as a unit of information. We develop all kinds of mental devices to help us group information into meaningful, or at least memorisable, chunks. We look for patterns in words and numbers. For example, remembering a mobile number is possible if we can see any patterns in the number sequence. A mathematical formula can hold several variables and can be seen as one chunk. Similarly, a chemical formula, for the experienced chemist, holds much information in one chunk. Indeed, a diagram or picture can often hold many items that are then seen as one chunk (Figure 1.4).

## 1.5 The Working Memory

Miller described the part of the brain as the '*short-term memory*'. He conceptualised this in terms of that part of the brain holding information temporarily before transfer to the long-term memory. This model developed from medical studies which had shown that, after a blow to the head, the patient often could remember nothing in the few minutes *preceding* the head injury but could remember more distant past events accurately. This led to the idea of the short-term memory and the long-term memory.

As more studies were conducted, it became apparent that this part of the brain not only could hold information on a temporary basis, but it was the part of the brain that *processed* the information. Indeed, it is now known that this is the *only* part where we process information, the long-term memory functioning as a store of almost infinite capacity. The name was changed to '*working memory*'. The phrase seems to have been used first by Massaro (1975). Many studies have confirmed that the capacity is fixed genetically. In fact, it can be shown that it grows to about age 16 when it remains fixed for the remainder of life, unless there is serious head injury or disease. Medical research gives evidence of its location, on two linked sites. Overviews in the context of science education can be found in Reid (2009a, 2009b).

The working memory is where we '*work mentally*': where we think, where we understand, where we solve problems. Information is taken in, held temporarily, worked on and, perhaps, understandings gained are stored in long-term memory or answers to problems written down on examination papers. The working memory can be cleared and made available for new information as needed. However, it has a fixed capacity and cannot be expanded.

There are two great differences between the '*expert*' and the '*novice*'. Greater experience enables the expert to group information together to be seen as one item, releasing more capacity for thought. Greater experience also allows the expert to focus only on what is essential for some task, leaving aside what is peripheral in the understanding of a concept or the solving of a problem (Figure 1.5).

In the early days, the phrase '*short-term memory*' was employed. Within a few decades, this started to be replaced by '*working memory*'. Figure 1.6 illustrates the distinction between the two descriptions: short-term memory and working memory.

From this point we shall refer to '*working memory*' as this description is more useful in thinking of the difficulties in understanding chemistry.

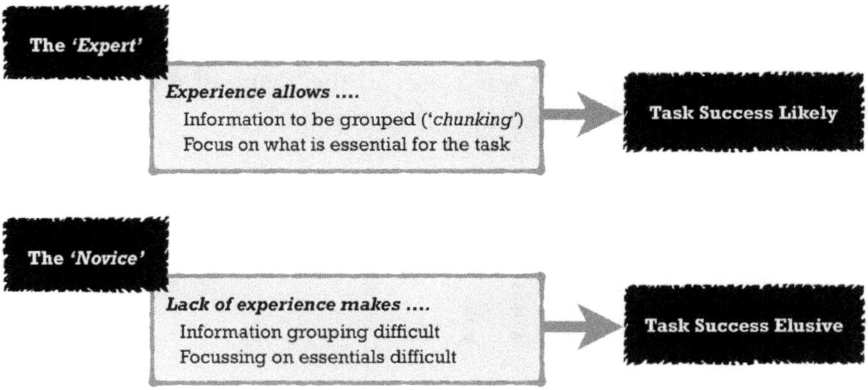


**Figure 1.5** 'Experts' and 'novices'.

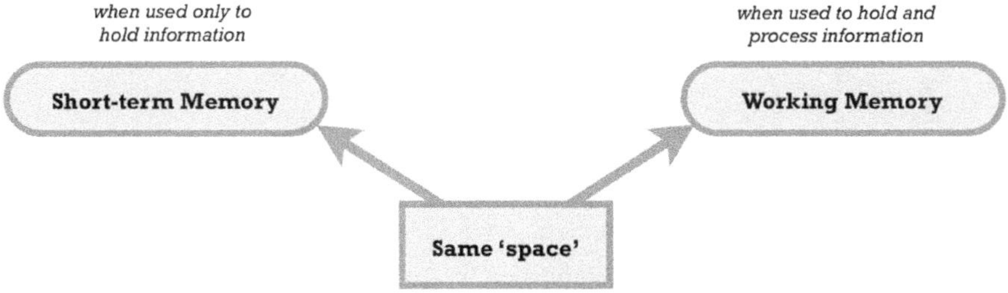


**Figure 1.6** Short-term memory and working memory.

Johnstone summarised much of this in his Brasted Lecture of 1996. This was published the following year (Johnstone, 1997).

## 1.6 The Breakthrough Experiment

Let us return to the work of Hamdi El-banna. His task was to test the insights and ideas that had been developed by Natalie Kellett (Johnstone and Kellett, 1980). In his research study, he started by gaining access to chemistry examination data for over 20 000 school students, aged about 16. Questions were considered in areas where there were known difficulties. Each question in the examination paper was analysed. Using a team of experienced teachers, the information load of these questions was determined. This load for each question was then checked by some students solving the questions out loud. They showed how they were approaching the question and could confirm whether the information load as seen by the experienced teachers was correct.

The average performance of the students in each question was plotted against the information load. It was expected that, as the load increased, average performance would fall. The work was carried out before the advent of the ubiquitous desktop and laptop computers of today. Therefore, in analysing the data, they used a main-frame computer and told the computer to draw the best-line-of-fit for the findings.

They observed that the best-line-of-fit seemed to miss quite a number of the points on the graph. It took some time before they realised that the relationship was not linear. It was closer to the shape of a titration curve. Figure 1.7 shows the kind of pattern they obtained.

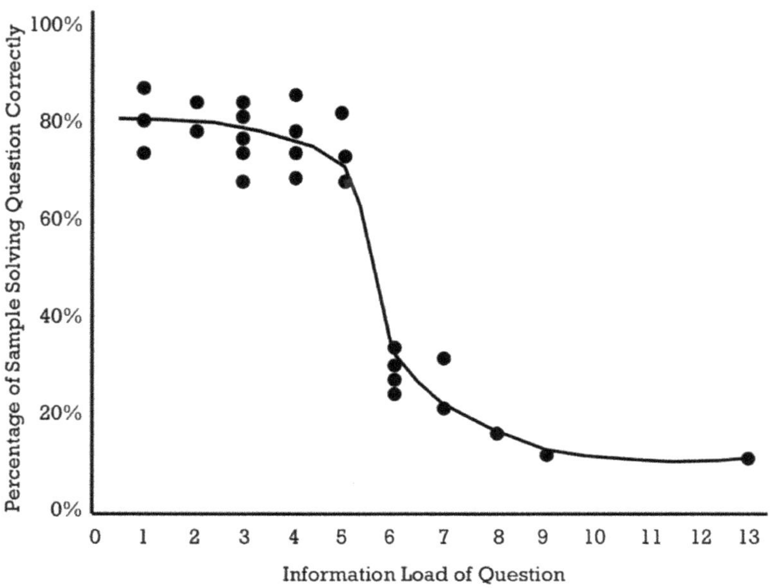


**Figure 1.7** Performance drop with working memory overload.

In the graph, we can see that the student success rate (% of the students who obtained the right answer) starts to drop cataclysmically when the information load reaches about six. In this, they thought of the information load as the sum of pieces of information in each question plus the additional pieces to be recalled plus the processing steps that were needed.

The next step that was then was to use the approaches developed by Miller and measure the capacity of the working memories of a sample of learners and then relate the measured capacity to performance in test items where the information load had been determined. This part of the study involved 300 first-year university chemistry students.

By this time, a number of ways of measuring the capacity of working memory were known and found to be valid and reliable. They employed two methods to measure the capacity of working memory of each student. In this, they sought to ensure that they were obtaining the most accurate measure. One is known as Digit Span Backwards Test (DSBT) and the other as the Figural Intersection Test (FIT).

***Digit Span Backwards Test***: learners have to report a list of numbers (of increasing length) backwards.
***Figure Intersection Test***: learners have to shade in overlaps in increasing numbers of overlapping geometrical shapes.

Only the students where the two methods of measurements gave the same outcome were then involved in the data analysis. However, this was the vast majority of their sample. The two methods are very different, the first numerical, the second geometric but they are known to give the same outcomes (within experimental error).

At the outset, they plotted performance against information load and gained a graph similar to that obtained with the examination data from the 20 000 school students (Figure 1.8).

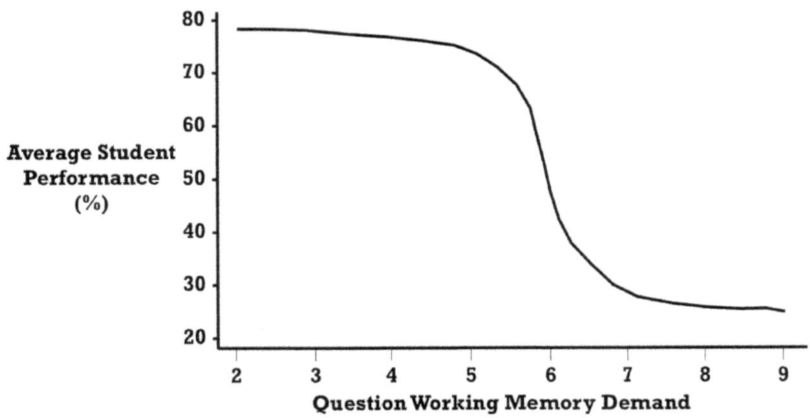


**Figure 1.8** Performance related to working memory demand.

As before, there was a cataclysmic collapse in performance when the load exceeds the capacity of the working memory. At that point, they brought in the data from the measured working memory capacity of the student group. They divided their student group into three groups, these groups being of similar size:

- Those with above average working memory capacity (>7).
- Those with average working memory capacity (7).
- Those with below average working memory capacity (<7).

It is interesting that they found that that almost every student had a measured capacity that lay between five and nine, with the average being seven. This is exactly what Miller found. Incidentally, it shows that the working memory capacity is not the same as intelligence or ability. It is a measure of the capacity of part of the brain. It seems that academic ability is related more to the *efficiency* by which we employ whatever capacity we have. Much is now known about this but, nonetheless, this is where much more research is needed.

For clarity, the graph they obtained is shown here somewhat more '*perfectly*' than their original (Figure 1.9).

What this graph shows is that it is the capacity of the working memory that *controls* success. The students with *above* average working memory capacities were successful until the information demand of the questions exceeds about six chunks. By contrast those with *below* average working memory capacities failed when the load was about two chunks less.

The findings were published in two papers (Johnstone and El-Banna, 1986, 1989). Later, Johnstone brought together the findings (Johnstone, 1991). The reaction of the readers was interesting. The outcomes were so precisely quantitative that they came as a surprise. However, the studies were

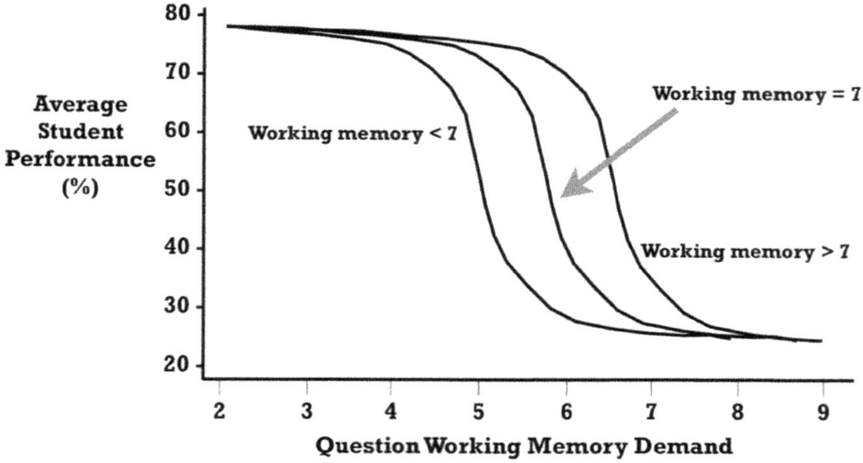


**Figure 1.9** Performance related to student working memory capacity.

repeated numerous time (for example: Johnstone *et al.*, 1993; Stamovlasis and Tsaparlis, 2000; Danili and Reid, 2004; Chen and Whitehead, 2009) and the same kind of pattern was observed. Therefore, one clear reason for difficulties in learning in chemistry (in fact, in all subject areas) is the fact that working memory capacity is limited and fixed.

There was another common reaction. Some looked at the range of working memory capacities: five to nine, with most possessing six, seven and eight. They then *assumed* that such variations were too small to be important. Numerous studies showed that this was simply not true. In terms of examination marks, the capacity of working memory can make a large difference in performance. Studies have shown that performance can typically rise by at least 15% (often much more) in an examination simply because the working memory is above average compared to below average.

Ali and Reid (2020) have brought together the outcomes of numerous studies. This is shown in Table 1.1. The final line in the table comes from a more recent study (Upahi and Ramnarain, 2020). The data are shown as correlation coefficients. Correlation shows the extent to which two variables (test performance and measured working memory capacities) are related. The data demonstrate the powerful influence of working memory capacity in gaining success in the typical kinds of assessment used today.

In Table 1.1, some studies used the Figural Intersection Test (FIT), some the Digit Span Backwards Test (DSBT). Many of these correlation coefficients are high, showing the impact of limited working memory capacity on test and examination performance. Of course, correlation does not, of itself, imply causation. However, the well-designed study of Johnstone and Elbanna (1986, 1989) demonstrates that the capacity working memory was *controlling* test and examination success.

In looking at the data in Table 1.1, it is not possible to make any deductions about difficult levels from the size of the correlation coefficient obtained. The value of the coefficient is influenced by the kinds of questions

**Table 1.1** Correlations: working memory capacity and performance, Source: Ali and Reid, 2020.

| Age | Country | Sample | Subject | Test used | Pearson correlation | Probability |
|---|---|---|---|---|---|---|
| 13–15 | India | 454 | Science | DSBT | 0.34 | $p<0.001$ |
| 13 | Kuwait | 641 | Six subjects | FIT | 0.22 | $p<0.001$ |
| 15 | Greece | 105 | Chemistry | FIT | 0.34 | $p<0.001$ |
| 13 | Taiwan | 151 | Physics | FIT | 0.30 | $p<0.001$ |
| 13 | Taiwan | 141 | Biology | FIT | 0.25 | $p<0.001$ |
| 13 | Taiwan | 141 | Genetics | FIT | 0.62 | $p<0.001$ |
| 6–17 | The Emirates | 809 | Physics | DSBT | 0.11 | $p<0.01$ |
| 16–17 | The Emirates | 349 | Physics | DSBT | 0.32 | $p<0.001$ |
| 16 | Greece | 90 | Mathematics | DSBT | 0.40 | $p<0.001$ |
| 11–12 | Pakistan | 150 | Mathematics | FIT | 0.69 | $p<0.001$ |
| 13–14 | Saudi Arabia | 800 | Sciences | FIT | 0.67 | $p<0.001$ |
| 18 | Nigeria | 665 | Chemistry | FIT | 0.81 | $p<0.001$ |

asked and the way they are framed. and does not just reflect the subject matter being tested. Good questions can be developed that are demanding but do not correlate with working memory capacity (Reid, 2002).

Overall, working memory capacity is one *major* factor controlling test and examination success. The capacity is not a proxy for ability or intelligence (whatever that is seen to be). The capacity each learner possesses is genetically determined and they have no control over it. What marks out the more able seems to be the *efficiency* by which each of us can employ whatever capacity we possess.

There are ethical issues here. Should learners be rewarded, at least in part, for the capacity of part of their brain over which they have no control. In assessment, we are trying to reward the achievements of learners, centrally based on assessing understanding. The danger is that we set questions that treat those with lower working memory capacities unfairly, not allowing them to show their true abilities, simply because their more limited working memory capacities cannot cope with the kind of questions we set.

If we look at the correlation data, the highest correlation between performance and measured working memory capacity obtained exceeds 0.8. Typical values in the maths-science areas seem to be nearer 0.4 although higher values are not uncommon. The variation in values is simply caused by the actual questions in the assessment. In some assessments, the question favour those with above average working memory capacities to a large degree, leaving those with below average working memory capacities seriously disadvantaged.

If we take the square of the correlation coefficient and convert to a percentage, it shows the percentage extent to which one variable relates to another. In the case, working memory capacity *controls* test success. Therefore:

$r = 0.8 \rightarrow 64\%$
$r = 0.7 \rightarrow 49\%$
$r = 0.6 \rightarrow 36\%$
$r = 0.5 \rightarrow 25\%$
$r = 0.4 \rightarrow 16\%$

Looking at this range of correlation coefficients, we can see that anything between 16% and 64% of the performance of the students is being *controlled* by the capacity of their working memories. Such correlations are quite typical in many examinations in the sciences and mathematics. This suggest strongly that we need to re-think the way we develop assessment questions.

## 1.7 Reasoning Chains

Let us stop for the moment and look at science teaching at primary stages as well as the first year or two at secondary levels. Johnstone discusses much loose thinking here (Johnstone, 1997). Just because we live in a '*scientific age*' (whatever that may mean) and are surrounded by gadgets and devices based on scientific research does not imply that science has to be taught to all. To gain enough insight to understand most of these is simply not possible in any school curriculum. Users of cars, washing machines, mobiles, adhesives and medicines do not have to understand the intricate details of the underpinning science.

It is much easier to justify science for all at school stages in terms of three broad goals (Figure 1.10). Every young person will move from formal education to play their part in wider society. They need to be equipped to understand something of the contributions that the sciences have brought to life, the implications of these developments and, very importantly, how the sciences have gained their insights which have, overall, often brought great benefits to humankind. Science is a human activity, undertaken by humans, working under the constraints of wider society.

At primary and early secondary stages, the working memory has not yet reached its full adult capacity: this is attained around the age of 16. Conceptual science demands, by its very nature, the holding of many ideas at the same time in order to gain understanding. Scientific thinking, by its very nature, means that we have to hold a series of reasoning steps *at the same time,* to gain insights. Until the working memories of learners have reached their full capacity, then it is simply folly to try to place much emphasis on conceptual science or on scientific thinking.

Consider the concept of energy. The word is overlain with all kinds of ideas gained from its use in wider society. If we then start to describe it in the formal terms of '*capacity to do work*', the young learner is immediately confused. The physics concept of '*work*' is not developed fully while the normal everyday use of the word carries somewhat different ideas. On top of that, the word '*energy*' conjures up all kinds of mental images: lacking

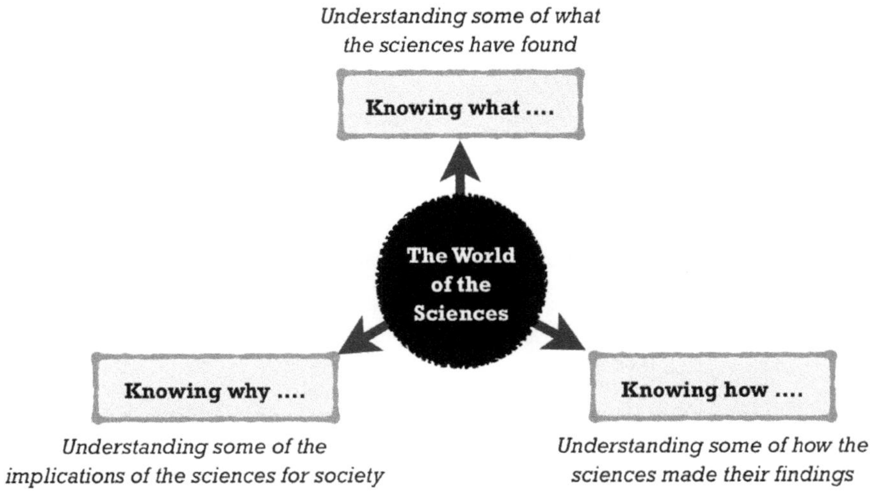


**Figure 1.10** Purposes for school science education.

energy, we want to flop down before a TV set. Then, we are informed by our media that there is an energy crisis. Then we hear that energy is conserved. If something is conserved, how can it run out? Confusion abounds.

Despite this, curriculum developers very often think it is a good idea to place it in curricula for primary classes. When chemistry starts to employ the idea, for example, in energy levels or bond energies, the learners have already accumulated a matrix of ideas, some which are contradictory. It seems essential to stress to curriculum designers that they must know and understanding the clear findings from psychology. In the science areas, the work of Johnstone, if applied, would resolve many of the problems which are often created by ill-informed curriculum designers. It is no use blaming the teachers: they do not have the control to solve problems not of their own making.

With colleagues, Johnstone looked at this by considering what he called '*reasoning chains*' (Johnstone *et al.*, 1997). One of his researchers explored the skills of primary school learners in employing reasoning chains and related this to measured working memory capacity, across various age groups. The data (from over 100 children across eight schools) generated a very instructive graph (Figure 1.11).

The graph is highly suggestive that limited working memory space is influencing the ability to follow reasoning chains. At this point, it is very instructive to watch skilled primary teachers working with children. This shows the way the teachers are able to prompt the children so that they can follow a longer reasoning chain.

The teacher, with a larger working memory and considerably more understanding and experience can hold the chain of ideas. The children need to be supported step by step to follow the logic and, unaided, are unable to sustain chains, except when very short. This again counsels caution in what we teach in the sciences at early stages.

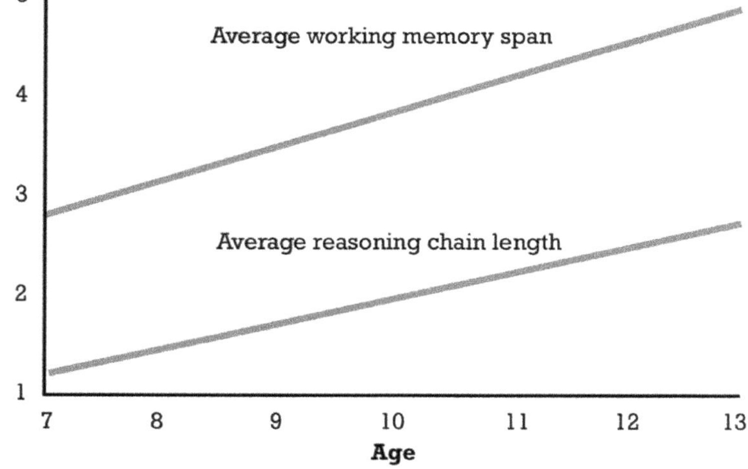


**Figure 1.11** Reasoning chains.

There are two fundamental principles for all science taught at primary stages:

(1) Avoid conceptual science.
(2) Embrace the descriptive—that which can be touched and seen.

Extremely valuable and useful chemistry can be taught without invoking the conceptual. We shall return to this later in this book, with illustrations that have been shown to work well. While these principles are critical at primary stages of learning, we must also be careful not to introduce the conceptual ideas of chemistry too early at secondary stages.

## 1.8 Wider Insights

Since the early 1970s, there has been a very large number of studies looking at working memory. Many have worked within the areas of science and mathematics education. It is in these subject areas where problems are greatest. Simply by their nature, understanding often requires the holding of many ideas at the same time. This places considerable stress on limited working memory resources.

It has been found that working memory controls extent of understanding in all subject areas. In her study of 13-year-olds, Hindal found that all six subjects taught at that age in her country generated significant positive correlations between working memory capacity and performance. However, the value for science tended to be higher (Hindal *et al.*, 2009, 2013; Hindal, 2014).

Working memory has also been studied from a psychological perspective. A leading figure here is Alan Baddeley (1934–) One of his major contributions is that he has now shown much more about the way the working memory functions. It is now apparent that working memory handles sound and vision in what he called '*loops*'. However, the working memory also has a management function. His long programme of research (Baddeley, 1986, 1994, 1996, 1997, 2000, 2002; Baddeley and Hitch, 1974; Baddeley *et al.*, 2015) has developed a picture where working memory can be seen as one entity, involved in four functions (Figure 1.12).

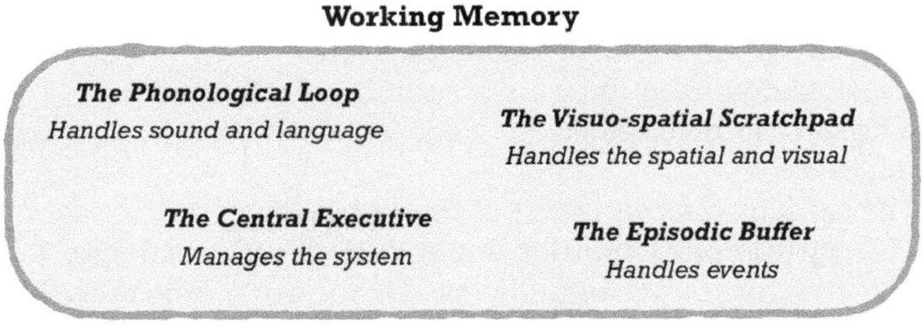


**Figure 1.12** Functions in working memory.

Working memory is known to grow with age. For example, while a 16-year-old has a capacity averaging seven (most ranging from five to nine), the average working memory capacity for a 12-year-old is nearer five. This immediately explains why certain topics and themes cannot be introduced too early in the education journey, something that every experienced teacher knows. It explains why a learner often cannot employ mathematical techniques when applied to chemistry or physics problems, much to the frustration of physics and chemistry teachers.

The working memory has difficulty in handling the mathematics as well as the concepts in the chemistry or physics. Once the mathematical procedures are '*automated*', they occupy very little space, leaving working memory capacity over to handle the chemistry and physics. Alenezi (2008) has found evidence that this probably offers the key to successful mathematics teaching and learning.

As we might expect, with lower working memory capacities at younger ages, primary education also has to take into account the importance of information overload. In a major text, Gathercole and Packiam (2008) have brought together the evidence relating to primary education and suggested some useful ways forward. Primary teachers are acutely aware that conceptual material is largely inaccessible to younger children. This has important implications for the sciences where any science taught at primary stages needs to be descriptive, avoiding conceptual thinking which, by its nature, make high demands on limited working memory capacity. This still applies in the early years of secondary education.

## 1.9 Key Messages

The rigorous research directed by Johnstone, drawing insights and findings from psychology, has shown that learning—when seen as understanding—is *controlled* by the capacity of working memory. Indeed, the working memory is that part of the brain where we carry out our thinking, understanding and solving of problems.

What is known now about the working memory can be summarised as:

**Working Memory**

- Brain location where we do our thinking, understanding and solving of problems.
- Capacity is fixed genetically: $7 \pm 2$.
- Capacity grows with age to about age 16.
- Capacity described in terms of '*chunks*' of information.
- A '*chunk*' is determined by what a person *perceives* as a unit of information.
- The capacity *controls* extent of understanding.
- The capacity often controls test and examination success.
- There are four functions undertaken in the working memory: handling sound, the visual and events as well as handling mental organisation.

One of the modern insights is that we can interpret and make sense of so many aspects of learning at all levels simply by considering the nature and function of the working memory. What Johnstone established nearly 40 years ago was the central place of limited working memory capacity in making sense of the difficulties our students often have when seeking to understand chemistry. What was more exciting was the way the insights his research revealed were able to help to predict what would make understanding more accessible. Much of this has been reviewed in Ali and Reid (2020).

## References

Alenezi D., (2008), *A study of learning mathematics related to some cognitive factors and to attitudes*, PhD thesis, Glasgow: University of Glasgow, http://theses.gla.ac.uk/333/, Accessed 7 November 2019.

Ali A. A. and Reid N., (2020), Making Sense of Learning, *A Research-based Approach*, Berlin: Springer (book currently in press).

Baddeley A. D., (1986), *Working Memory*, Oxford: Oxford University Press.

Baddeley A. D., (1994), The Magical Number Seven: Still Magic after all these Years? *Psychol. Bull.*, **101**(2), 353–356.

Baddeley A. D., (1996), Exploring the central executive, *Q. J. Exp. Psychol.*, **49A**(1), 5–28.

Baddeley A. D., (1997), *Human Memory: Theory and Practice*, Hove: Psychology Press Ltd.

Baddeley A. D., (2000), The episodic buffer: a new component of the working memory? *Trends Cognit. Sci.*, **4**(11), 417–423.

Baddeley A. D., (2002), Is Working Memory Still Working, *Eur Psychol.*, 7(2), 85–97.

Baddeley A. D. and Hitch G. J., (1974), Working memory, *Recent Advances in Learning and Motivation*, vol. 8, pp. 47–89.

Baddeley A. D., Eysenck M. W. and Anderson M. C., (2015), *Memory*, 2nd edn, Glasgow: Psychology Press.

Chen W.-C. and Whitehead R., (2009), Understanding physics in relation to working memory, *Res. Sci. Technol. Educ.*, **27**(2), 151–160.

Curriculum Papers 512, (1962), *Alternative Chemistry for Ordinary and Higher Grade*, Dalkeith, Edinburgh: Scottish Examination Board.

Danili E. and Reid N., (2004), Some Strategies to improve performance in school chemistry, based on two cognitive factors, *Res. Sci. Technol. Educ.*, **22**(2), 203–226.

Gathercole S. E. and Packiam T. P., (2008), *Working Memory and Learning: A Practical Guide for Teachers*, London: Sage Publications.

Hindal H., (2014), *Cognitive Characteristics of High Achievers*, Germany: Lambert Academic Publishing.

Hindal H., Reid N. and Badgaish M., (2009), Working Memory, Performance and Learner Characteristics, *Res. Sci. Technol. Educ.*, **27**(2), 187–204.

Hindal H., Reid N. and Whitehead R., (2013), A Fresh look at High Ability, *Int. J. Instr.*, **6**(1), 59–76.

Johnstone A. H., (1991), Why is Science Difficult to Learn? Things are Seldom What They Seem, *J. Comput. Assist. Learn.*, **7**, 75–83.
Johnstone A. H., (1993), The Development of Chemistry Teaching, *J. Chem. Educ.*, **70**(9), 701–705.
Johnstone A. H., (1997), Chemistry Teaching, Science or Alchemy? *J. Chem. Educ.*, **74**(3), 262–268.
Johnstone A. H. and El-Banna H., (1986), Capacities, Demands and Processes: a Predictive Model for Science Education, *Educ. Chem.*, **23**(3), 80–84.
Johnstone A. H. and El-Banna H., (1989), Understanding Learning Difficulties - A Predictive Research Model, *Stud. High. Educ.*, **14**(2), 159–168.
Johnstone A. H. and Kellett N. C., (1980), Learning Difficulties in School Science - Towards a Working Hypothesis, *Eur. J. Sci. Educ.*, **2**(2), 175–181.
Johnstone A. H. and Morrison T. I., (1964–1969), *Chemistry Takes Shape*, London: Heinemann Educational, vol. 1–5.
Johnstone A. H., Hogg W. R. and Ziane M., (1993), A working memory model applied to physics problem solving, *Int. J. Sci. Educ.*, **15**(6), 663–672.
Johnstone A. H., Hogg W. R., MacGuire P. R. P. and Raja S. H., (1997), How long is a chain? Reasoning in Science, *Sch. Sci. Rev.*, **78**(285), 3–77.
Johnstone A. H., Morrison T. I. and Sharp D. W. A., (1971), Topic Difficulties in Chemistry, *Educ. Chem.*, **8**, 212.
Kellett N. C., (1978), *Studies on the perception of organic chemical structures*, PhD thesis, Glasgow: University of Glasgow.
Massaro D. W., (1975), *Experimental Psychology and Information Processing*, Chicago: Rand McNally.
Miller G. A., (1956a), The magical number seven, plus or minus two: some limits on our capacity for processing information, *Psychol. Rev.*, **63**, 81–87.
Miller G. A., (1956b), Information and memory, *Sci. Am.*, **195**(2), 42–46.
Reid P. A., (2002), *Problem solving by primary school children with particular reference to dyslexics*, MSc thesis, Glasgow: University of Glasgow.
Reid N., (2009a), The Concept of Working Memory, *Res. Sci. Technol. Educ.*, **27**(2), 131–138.
Reid N., (2009b), Working Memory and Science Education, *Res. Sci. Technol. Educ.*, **27**(2), 245–250.
Reid N. and Skryabina E., (2002a), Attitudes Towards Physics, *Res. Sci. Technol. Educ.*, **20**(1), 67–81.
Reid N. and Skryabina E., (2002b), Gender and Physics, *Int. J. Sci. Educ.*, **25**(4), 509–536.
Scottish Qualifications Authority, (1998), *Comparability Study of Scottish Qualifications and GCE Advanced Levels, General report and Subject Reports*, Dalkeith: Scottish Qualifications Authority.
Stamovlasis D. and Tsaparlis G., (2000), Non-linear analysis of the effect of working memory capacity on organic synthesis problem solving, *Chem. Educ. Res. Pract.*, **1**(3), 375–380.
Upahi J. E. and Ramnarain U., (2020), Examining the connection between students' working memory and their abilities to solve open-ended chemistry problems, *J. Balt. Sci. Educ.*, **19**(1), 142–156.

CHAPTER 2

# The Central Role of Working Memory: How We Handle Information

## 2.1 Looking Back

In the last chapter, we saw how Johnstone was seeking to address the fundamental question: why is it that understanding chemistry is so often found to be demanding? He wondered if it was something about the nature of chemistry or something about how humans gain understanding. He started in the 1960s by identifying the areas where the difficulties were greatest. In the process, he developed a very straightforward but valid and reliable way of pinpointing the areas of greatest difficulty. The approach he adopted has been taken up by others in other subject areas. Thus, Ali employed this successfully in mathematics learning in a developing country (Ali and Reid, 2012).

His next step was to explore many of these areas to see why difficulties were occurring and, more importantly, to explore ways by which the difficulties might be reduced. Many insights were gained and several very useful developments emerged, but it took a long time before the work of Kellett gave him the breakthrough he was seeking (Johnstone and Kellett, 1980).

## 2.2 Other Insights

At that time, the fashion in much work was to look at what became known as misconceptions as well as alternative conceptions. The former are understandings that are wrong in some way while the latter presented understandings that did not fit with the prevailing understandings. The numbers of studies in these areas was enormous. Two reviews reveal just how much work

Advances in Chemistry Education Series No. 6
The Johnstone Triangle: The Key to Understanding Chemistry
By Norman Reid

Published by the Royal Society of Chemistry, www.rsc.org

was carried out although, sadly, little emerged that helped to address the difficulties learners were experiencing (Ozmen, 2004; Hodgson and Pyle, 2010).

Much was driven by the idea of '*constructivism*'. The central idea here is that '*learners construct their own understandings and that these may or may not be the same as the understandings intended by the teacher*' (Driver and Oldham, 1986, p. 105). Osborne and Wittrock (1983, p. 492) observed wisely that, '*The brain is not a passive consumer of information. Instead it constructs its own interpretations of information and draws inferences from them*'. The central idea is simple, but education has often moved off in multiple directions, frequently leaving a considerable level of confusion (Geelan, 1997).

The science education world readily embraced the ideas of constructivism. Sadly, this led to few insights that enabled learners to gain enhanced understanding or perform better in tests and examinations. In her review, Solomon (1994) described the trends in terms of the '*Rise and fall of constructivism*'. Much later Scerri (2010, p. 17) was even more critical when he talked of the '*fading claims that constructivism is a viable approach in science education*'.

Johnstone had the insight to foresee that constructivism would not help us *predictively* in our attempts to make chemistry more accessible for learners. Therefore, he rejected the pursuit of any constructivist agenda. This was not because the central tenet of constructivism is incorrect but for the practical reason that constructivism is an excellent *description* of understanding (what goes on in the working memory of the learner), but it is *predictively* not useful at all when comes to selecting teaching approaches. Figure 2.1 illustrates this central principle.

In this, we must emphasise that learners construct *understandings*. The process is natural, and it occurs *within* the working memory. Learners do not construct knowledge. Knowledge is there whether the learners know it or not. They are seeking to make sense of what is presented to them, formally in the school or university context or informally through wider life experiences (Figure 2.2).

We need to stop and think about how we understand the concept of '*understanding*'. There is a danger here. We are *constructing our understanding of understanding*! However, we may not share exactly the same understanding of

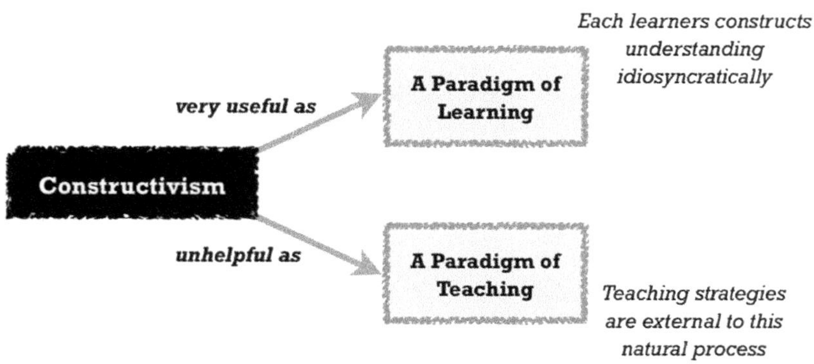


**Figure 2.1** The paradigm of constructivism.

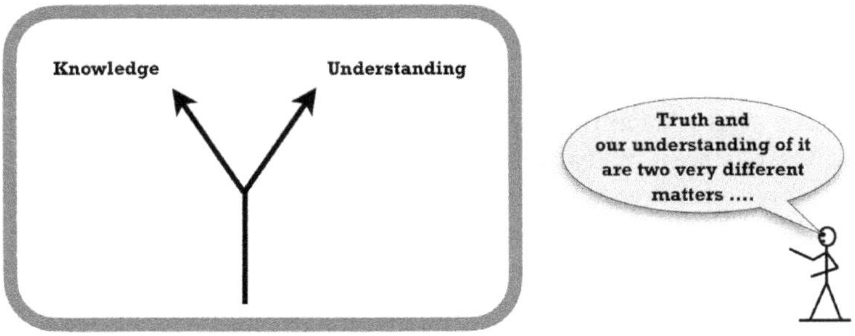


**Figure 2.2** Knowledge and understanding.

what we mean by understanding and this is often where educational texts can lead to considerable confusion. Some use the word '*knowledge*' when the author clearly means '*understanding*'. We are also influenced by the Bloom's taxonomy where we often see the idea of '*comprehension*' as equating to understanding and, therefore, separate from '*knowledge*' or '*application*' (Bloom, 1956).

In a heavily referenced review of assessment written for an Examination Board in a developing country, these issues were analysed (Ud-Din *et al.*, 2016). This led to a description of *understanding* which was:

**Understanding**
When a person really understands an idea, then the individual is able to apply that idea in a novel situation with a good prospect of success.

This was developed in order to clarify goals in an assessment context. In essence, what it is saying is that the *evidence* for understanding can be seen when ideas are applied successfully. The power of this in an assessment context is readily apparent.

In his earlier school career, Johnstone had been the Principal Examiner for Chemistry at all levels under the then Scottish Examination Board (now re-named as the Scottish Qualifications Authority). While Johnstone never specified a description of understanding in the way shown above, this idea underpinned how questions that tested understanding were being developed. Indeed, some of the questions in Chemistry papers during several years in the late 1960s and early 1970s (when he was principal examiner) are pioneers in the development of exciting new ways to assess chemistry.

## 2.3 Johnstone's Central Question

Let us return to the central issues that underpinned the research of Johnstone throughout much of his career: why is understanding chemistry often demanding. He had realised that there were two sub-questions: was there something about the nature of chemistry or something about how humans gain understanding. Establishing the central role of working memory started to answer the second question. However, before we turn to the first question, we need to see how his research developed in the 1980s and 1990s.

He was developing research projects in two broad directions. One related to attitudes where he appreciated that there might be connections between demand level as seen by learners and the attitudes they developed. A series of studies offered some very useful insights but, incidentally, started to undermine any idea that perceived difficulty was neatly related to the generation of more negative attitudes (Hadden and Johnstone, 1982, 1983a, 1983b). Later work in physics demonstrated that school learners were developing extremely positive attitudes towards physics, despite the finding that the curriculum they were following was exceptionally demanding (Reid and Skryabina, 2002). Indeed, the Scottish curriculum in Physics was so demanding that a rigorous study showed that an A pass in the Scottish Higher Grade at that time was roughly equivalent to an A pass in the English A level, despite the Higher being sat one year earlier and involving one year less study (SQA, 1998). Demand level and enjoyment are not neatly related as Yang later confirmed in a very different context, this time in chemistry (Yang, 2000; Reid and Yang, 2002b).

Demand level and motivation to pursue further studies are also not neatly related. The key factor that enhances motivation relates to extent of understanding. If working memory is unable to cope, understanding is the casualty. To pass examinations, the learners have to resort to memorisation. This is not the natural way to learn as Piaget revealed long ago when he observed that learners were trying to make sense of what was presented to them (Flavell, 1963; Atkinson, 1983; Bliss, 1995; Wadsworth, 2004). When learners have to resort to memorisation, then learners do not experience the satisfaction of being able to make sense of what is given to them. It is this lack of satisfaction that underpins less positive attitudes and falling motivation (Figure 2.3).

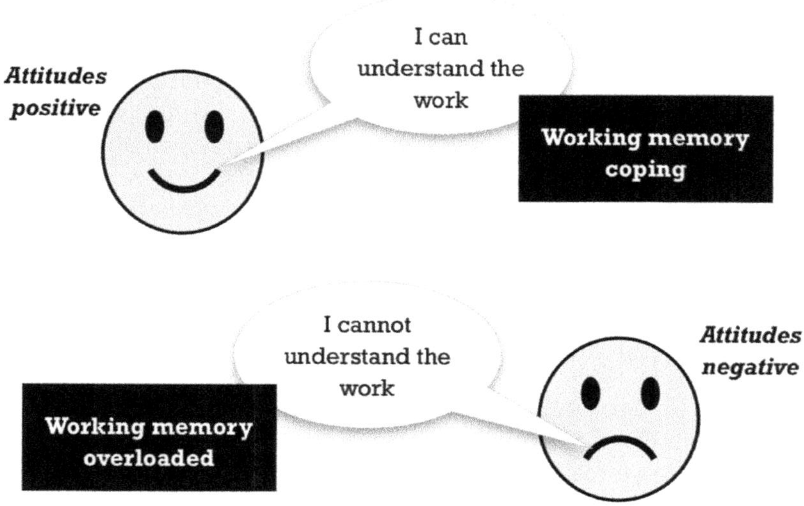


**Figure 2.3** Keys to positive motivation.

Later work demonstrated the central place of understanding in generating positive attitudes (Jung, 2005; Hussein and Reid, 2009; Jung and Reid, 2009), but the findings have rarely been applied anywhere in the world. The key important message is enhance understanding by presenting chemistry in such a way that working memory is less likely to be overloaded. Positive attitudes and motivation will be enhanced.

**Attitudes and Motivation**
The simple principle is that when learners are able to understand what is presented their attitudes develop positively. Understanding is controlled by limited working memory capacity. Therefore, when we teach in line with the limitations of working memory capacity, then we gain positively motivated students. Learners at all ages '*see through*' being presented with undemanding material. They want challenge but understanding must not be made impossible.

One of the sad features of much educational research work today is the excessive use of questionnaires. Many studies have addressed the problem of poor attitudes towards the sciences (and hence deteriorating motivation to study them). They have started to approach this by developing all kinds of questionnaires to measure attitudes. The simple fact is that we cannot measure attitudes towards to sciences with sufficient precision using questionnaires to give any really helpful insights. The literature relating to the limitations in using questionnaires has been reviewed some detail in Ali and Reid (2020). This brings together major issues identified in Reid (2006, 2015) where the incorrect way of handling of questionnaire data is discussed.

The tendency is to start with motivation and end up encouraging teachers to develop better motivation among their students. However, the key problem here is that we are starting at the wrong end. Positive motivation is *generated* by successful understanding (meaningful learning). Therefore, we need to start by focussing on how to enhance understanding that is meaningful for the learners, not on the motivation.

In fact, we have seen that the answer already exists in the literature. If we develop curricula as well as teach and assess in such a way that the possibility of working memory overload is reduced, this will generate better understanding. When the learners perceive that they do understand what is being presented to them, their attitudes towards their studies rise. This will be seen in due course in increasing numbers opting to stay on in studying chemistry. That holds one important key when there is any problem in the numbers opting to study chemistry.

**Popularity of Chemistry**
Two factors have been found in several studies:

1. Develop curricula that enable learners to make sense of their world from a chemistry perspective.

2. Develop and teach the chemistry curriculum in such a way that working memory is less likely to be overloaded.

When the research evidence is brought together, there is a sequence of thinking and this can be represented as in Figure 2.4.

We shall develop this further, with examples, later. The key thing to note, for the moment, is that far too many studies have preceded in the wrong direction. We cannot try to measure motivation. Indeed, it is very difficult to measure (for it is multivariate). It is, therefore, pointless to blame teachers for lack of learner motivation, suggesting that they do something about this. Instead of starting with motivation, the key is to start by focussing on enhancing understanding and the key to that lies in the working memory.

## 2.4 Further Insights

In the light of the findings of Kellett and El-Banna, Johnstone set about reading around to see if there were any useful insights coming from psychology that might take his understanding forward. The findings related to working memory were starting to be taken up in psychology on a large scale and many started to develop models of the way information is handled in the human brain in the light of the evidence that was being found.

As early as 1960, Miller and his colleagues realised that what had been described as '*short-term memory*' not only functioned as a short-term

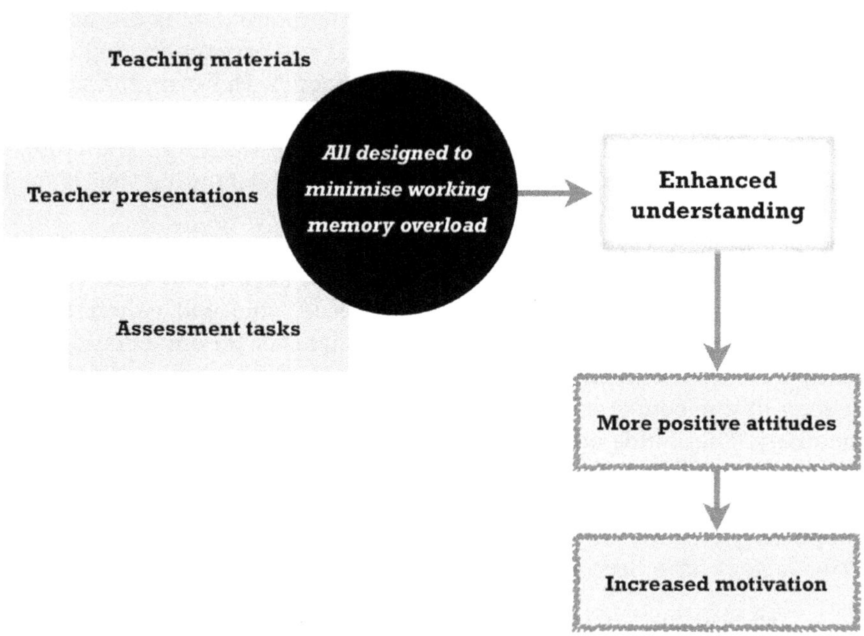


**Figure 2.4** From Understanding to Motivation.

memory store but also carried out the mental processing required in thinking, understanding and problem solving (Miller *et al.*, 1960). Atkinson and Shiffrin (1968) took this forward in a very influential chapter where they developed a multicomponent model of human memory. At various times, numerous reviews appeared in the literature and these have brought together the evidence in helpful ways (Baddeley and Hitch, 1974; Miyake and Shah, 1999; Baddeley, 2007). However, there are two very useful more modern reviews that bring together considerable evidence in accessible ways: Baddeley (2010) and Sweller (2011).

The way Atkinson and Shiffrin (1968) saw working memory was that it was one '*unitary space*' in the human brain. The sustained research of Baddeley (supported by other studies) has now shown that the place can be seen as holding four functions (Baddeley *et al.*, 2015). Many useful insights have also been gained from clinical psychology.

Nonetheless, in science education, the dominant influence has been to consider working memory as one space whose capacity controlled understanding and assessment success. This reflects the central importance in science education in seeking ways to help learners gain greater understanding in highly conceptual areas as well as the pressing need for these learners to perform well in examinations. Nonetheless, there is a strong argument that suggested that we need to look in much more detail at the functions in the context of science education (St Clair-Thompson and Botton, 2009; St Clair-Thompson *et al.*, 2010).

In addition, the development of fresh insights on the importance of the visual-spatial has also raised questions about the role of the visual-spatial in learning: is this a feature of the way the working memory operates or is it a characteristic of how understandings are stored in long-term memory? The work of Hindal has raised some fascinating questions about the nature and role of the visual-spatial in learning (Hindal, 2007; Hindal *et al.*, 2009; Hindal *et al.*, 2013; Hindal, 2014). Working with 13-year-old learners, Hindal developed a computer-driven test to measure the visual-spatial characteristic. In the process, she uncovered a considerable amount of research evidence that confirmed the central importance of the visual-spatial in all learning (Silverman, 1989, 2002).

The visual-spatial learner tends to see things in terms of relationships rather than in a logical series of arguments or stages of thought. It appears that a high proportion of the population naturally thinks this way. Education, over the decades, has tended to develop systematic ways of presenting material, following a logical process of thought. This disadvantages many and it was fascinating to see that Hindal found that those who were more able to think in a visual-spatial way performed very much better in examinations across six different subject areas (one of which was science) in a system were the rewards come largely from recall. This finding was confirmed several times. This has major implications for a subject like chemistry which is so full of concepts and where these concepts relate to each other in quite complex ways.

Given the evidence that those who demonstrated high levels of visual-spatial thinking seemed to perform very much better than others, this raises interesting questions about the nature of human ability. What Hindal was seeing was that those who were very gifted academically tended also to rely on visual-spatial thinking: seeing things in terms of relationships between ideas. Historically, Albert Einstein (1879–1955) thought in this way and, although he never performed well at school, he made some of the greatest contributions in physics and astronomy. However, it not just in the area of the sciences where this pattern is to be seen. Wolfgang Amadeus Mozart (1756–1791) saw the '*shape*' of the music that he was composing: perhaps the visual shape of the music was more important to him than logical notation. His creative musical output during a very short life was prodigious.

The importance of working memory in learning has been understood in the world of psychology and educational psychology. One of the sad aspects of wider education literature is the way educational texts have, in general, failed to emphasise the central importance of working memory capacity as holding the key to understanding. This is holding back teacher education, curriculum construction and assessment development. This is where the insights from Johnstone are so important. He was one of the first to bring the research evidence across into education and he demonstrated the critical importance of working memory capacity in understanding and in assessment.

A recent review offers a rigorous but accessible overview of the research evidence, with a strong educational slant, as well as providing a rich range of references (Cowan, 2014). Cowan makes the central point when he states:

> *Given the ability of humans to use long-term memory so adeptly, one could ask why we care about the severe working memory capacity limit at all. The answer is that it is critical when there is limited long-term knowledge of the topic. In such circumstances, the capacity of working memory can determine how many items can be held in mind at once in order to use the items together, or to link them to form a new concept in long-term memory. This is the case in many situations that are important for learning and comprehension* (Cowan, 2014, p. 205).

In their review, Kirschner *et al.* (2006, p. 76) make the same point eloquently when they state, '*Any instructional procedure that ignores the structures that constitute human cognitive architecture is not likely to be effective*'. What they are saying links back to the much earlier work of David Ausubel (1968). What we already understand and is held in long-term memory gives us the ability to chunk new material so that we can make sense of it. If we do not possess the necessary background understandings, then chunking is more difficult, the working memory overloads and understanding of new material becomes almost impossible.

Aware of this, Cowan advises: '*To avoid overtaxing an individual's working-memory capabilities, one should avoid presenting more than a few items or ideas at once, unless the items can be rapidly integrated*' (Cowan, 2014, p. 206). This gives us practical advice: in teaching we need to overtly link new ideas back to ideas that are established in the minds of the learners while we need to introduce new ideas step-by-step. All this is critical in the light of the clear research evidence about how humans gain understandings. What we already understand is allows us to make sense of what is new, the working memory being central in gaining understanding.

The limitations in the capacity of working memory also link back to the findings of Jean Piaget (1896–1980). Piaget is well known for describing a series of developmental stages which he had observed with young children. He offered no underlying reason. Piaget's students later started to explore possible reasons for what was observed. They started to see that the developmental increases could be related to increases in the number of items that could be held mentally *at the same time* (Pascual-Leone and Smith, 1969). We are back to the limited capacity of working memory and the observation that this grows with age to about age 16.

Cowan looks at the role of the working memory in concept development. He notes the central importance of ideas and information being brought together mentally. This takes place in the working memory and he summarises the process: '*For the various types of concept formation, then, the cauldron is assumed to be working memory*' (Cowan, 2014, p. 210). This insight holds the key to effective teaching and learning.

Halford *et al.* (1998) stress that solving a problem requires a satisfactory combination of working memory capacity, speed of thinking, background knowledge and understanding, as well as having suitable strategies available. This links closely to the findings of Yang (2000) where her focus was on the solving of the more open-ended problems in chemistry at school level. She summarised the literature related to problem-solving in general (Reid and Yang, 2002a) and, in a second paper, outlined what she found when teenage students worked in groups to solve open-ended chemistry problems. In this, she was focussing on the long-term memory in that the importance of working memory had already been established (Reid and Yang, 2002b).

When faced with an open-ended chemistry problem, the evidence shows that students look for *anything* they recognise that might just possibly be helpful. The working memory is searching the long-term memory for any understandings that relate to the problem. It appears that one major factor is the way ideas and understandings are linked when they are stored in the long-term memory. Confirmed by much other research, this suggests that the successful problem-solver (in open-ended problems) is the person who not only possesses the appropriate understandings that are needed but can access them in a useful way. Such access arises when key ideas are well linked in the way they are stored. The original storage and linking depends on the functioning of working memory in their past.

In the light of the research evidence, Cowan (2014) emphasised the importance of presenting ideas step by step, allowing time for ideas to be integrated mentally. This reduces the potential for working memory overload. He also stresses the importance of not introducing ideas to be introduced too early. We need to wait until the background understanding is secure as well as allowing for the working memory capacity to develop with age. This approach makes sense of the idea of the spiral curriculum advocated by Bruner (1960). A topic is introduced giving only the key ideas. Later on (sometimes even a year after), learners return to the topic in greater depth. This process can happen three or four times and mature understandings can then be developed.

This was a feature of the Scottish Chemistry curriculum at one time. For example, very simple volumetric calculations were touched upon at about age 15. There was a return to the topic one year later, when the calculations were still acid–base but more demanding and extending into the gravimetric. One year later again, the ideas were extended into redox volumetric and even complexometric analysis. A very powerful way was developed to aid the working memory to enable students to gain success (Reid, 2010). We shall return to that later.

Understanding the central role of working memory also has implications for curriculum design, an issue that Johnstone (2000) addressed from a different perspective. The work of Sweller and his colleagues is important in this context. They stress the, '*central importance of working memory constraints as determinants of instructional design effectiveness*' (Sweller *et al.*, 1998, p. 251). Here, there are some subtle but very important insights. In essence, if the cognitive load is too high, understanding will be the casualty as the working memory is unable to cope. However, if the cognitive load is too low, there is little challenge and the learners may disengage from the task. Learners thrive on challenge, but the challenge must not be such that the working memory has no chance of coping.

**Observation**

It is often observed that the curriculum demands in the first two years in secondary schools in Scotland are set far too low. This brings no real advantages to the slower learners while the faster learners are bored, with no challenge. There is a need to set the challenge level correctly. The evidence shows that all learners like challenge and the feeling that they have had some success in meeting the challenge. School learners hold little respect for being forced to learn trivia.

There are important issues when multimedia and visual presentations are employed. The working memory has what Baddeley called '*loops*': a visual-spatial loop and an auditory loop (Baddeley *et al.*, 2015). Much more work needs to be carried out looking at the balance of these. Again, it seems that, if too much is presented verbally, then little capacity is left for the visual and *vice versa*. It is the overall load that seems critical. However, the visual has the potential in grouping ideas together and the working memory '*sees*' a group of ideas as one. This may offer some kind of interpretation of what Hindal found when she observed that those who were strong in the visual–spatial always performs better in tests and examinations (Hindal, 2007, 2014). It seems that the visual–spatial is being employed as a chunking device.

There is also another issue—that of language. Cowan (2014) notes that, '*the working memory demands of language production do not always match the demands of the recipients' language comprehension*'. Earlier, Johnstone had found that, when learners were working in their second language (even when they were competent at that language), then the measured working memory fell by about one chunk (around 15%) (Johnstone and Selepeng, 2001). This suggests that part of the working memory capacity was being employed for some kind of '*interpretation*'. This implies that working in a second language reduces the usable working memory by about 15%, placing those who have to work in a second language at a considerable disadvantage.

This has important implications for many learners of chemistry worldwide. Much chemistry is taught in English in India, Pakistan, the Middle East and North Africa where English is not the native language. In addition, the languages used in these countries write from right to left, using a different script. The places learners at a working memory disadvantage when learning and, even more importantly, when being assessed in formal examinations using the English language. We can go further. Imagine what it must feel like for a Japanese or Chinese student when they first meet the Periodic Table!

While the evidence shows quite clearly that the capacity of working memory for any individual is determined genetically and that the capacity does not relate neatly to ability or intelligence, many have attempted to increase the capacity, or perhaps increase the efficiency of that capacity. It is not easy to test claims of success, but it does seem that increasing efficiency may be possible but increasing capacity is impossible.

The central issue is that humans are pattern-seekers and this pattern-seeking helps us to '*chunk*' ideas and use our working memories more efficiently. Training programmes are almost certainly allowing chunking strategies to be practised. However, there is no evidence to suggest that the training in any one area transfers to another. Therefore, what is being done is that, by means of practice, we are giving the learners some strategies to improve performance. However, these strategies may be specific to the themes involved in the training, little transferring to other tasks.

## 2.5 Information Processing

In the 1960s, the central importance of working memory in all learning was beginning to be appreciated in the psychology of learning. In their seminal contribution, Atkinson and Shiffrin (1968) had brought together much useful evidence. What was known was that the brain operated with two memories: the working memory and the long-term memory. In some way, new information and ideas came into the working memory where they were processed before, perhaps, the outcomes being stored in the long-term memory. This generated a flow model (Figure 2.5).

Johnstone took this model and combined findings from other research, mainly in the psychology research literature, to develop his now well-established information processing model. This can be seen in various

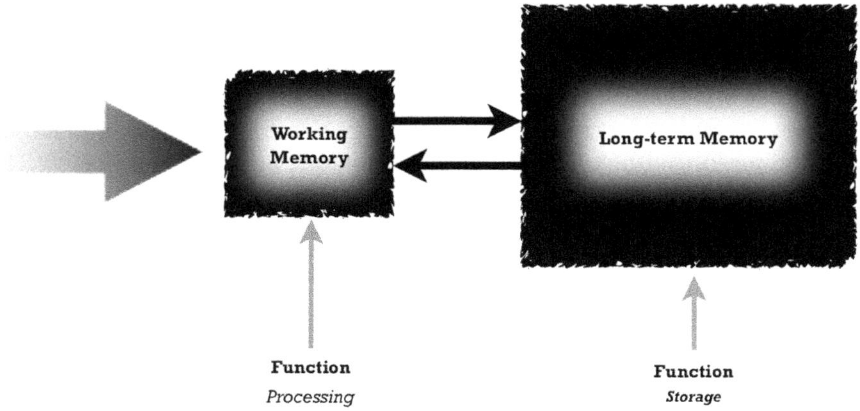


**Figure 2.5** Two memories.

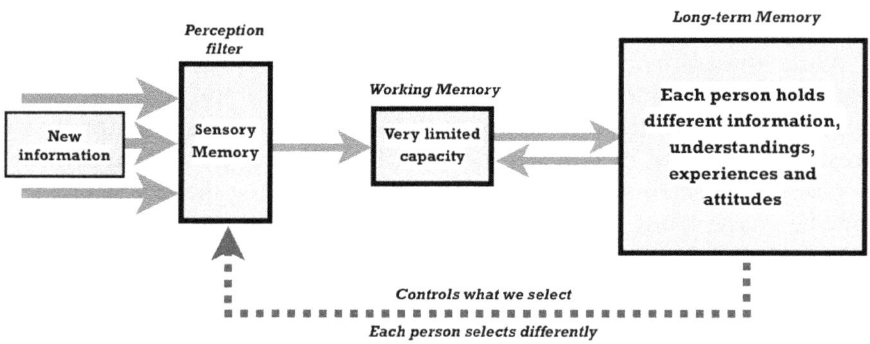


**Figure 2.6** Johnstone's model.

forms in Johnstone (1991, 1993, 1997). This was taken up enthusiastically by university chemistry teachers who quickly grasped the way the model not only interpreted so much evidence that they were observing in their teaching but also provided predictive ways forward to guide into strategies that were likely to enhance learning when seen an understanding. The author remembers one university chemistry education research conference where five out of the six presentations were based on the model, each showing Johnstone's model in screen, before showing how their research was built on predictions from the model. Figure 2.6 shows the model that Johnstone developed.

Let us consider a teaching situation where learners are being presented with chemistry information. The model can interpret what is going on. The information enters mainly in the form of sound and vision and it goes into what is known as a sensory memory (Johnstone later re-named this as a perception filter). This sensory memory selects what is to be passed on to the working memory. The selection is controlled by what the learner already knows, understands and has experienced. This is stored already in long-term memory. David Ausubel (1918–2008) had demonstrated the central importance of previous knowledge and understanding controlling the extent of new understanding with new material (Ausubel, 1968).

> *If I had to reduce all of educational psychology to just one principle, I would say this: the most important single factor influencing learning is what the learner already knows. Ascertain this and teach him accordingly.*
>
> Ausubel, 1968, preface

The working memory then receives what is passed to it and will draw from the long-term memory anything that seems useful so that the new input makes some kind of sense. At that stage, new understandings or information can be passed to the long-term memory for storage.

Johnstone (1997, pp. 264–265) looked at ways in which new information or understandings can be stored and suggested the following four situations:

(1) *The new knowledge finds a good fit to existing knowledge and is merged to enrich the existing knowledge and understanding (correctly filed).*
(2) *The new knowledge seems to find a good fit (or at least a reasonable fit) with existing knowledge and is attached and stored, but this may, in fact, be a misfit (a misfiling).*
(3) *Storage can often have a linear sequence built into it, and that may be the sequence in which things are taught.*
(4) *The last type of memorisation is that which occurs when the learner can find no connection on which to attach the new knowledge.*

The first way gives enhanced understanding, what Ausubel (1968) described as *meaningful learning*. The second way generates misconceptions (or perhaps alternative conceptions). The third relates to things like learning the alphabet, memorising times-tables or spelling in primary education. The fourth is simply rote memorisations which leads to recall to pass examinations. Indeed, it is questionable if this kind of rote memorisation (where there is no meaning for the learner) has any place in genuine education.

Let us look again at the esterification reaction we considered in Chapter 1.

$$C_2H_5OH + CH_3COOH \rightarrow CH_3COOC_2H_5 + H_2O$$

The '*novice*' learner may lack enough previous knowledge and understandings to make sense of the reactant formulae and, with the needs of examinations looming, may attempt to memorise the equation and store it this way. However, another learner may understand what the various formulae represent and see the equation as a way to summarise what was observed in the laboratory (meaningful learning). However, yet another learner may link up the idea that the reaction is like a neutralisation. The ester is seen as rather like a salt while the ethanol behaves like an alkali. This has the makings of several potential misconceptions.

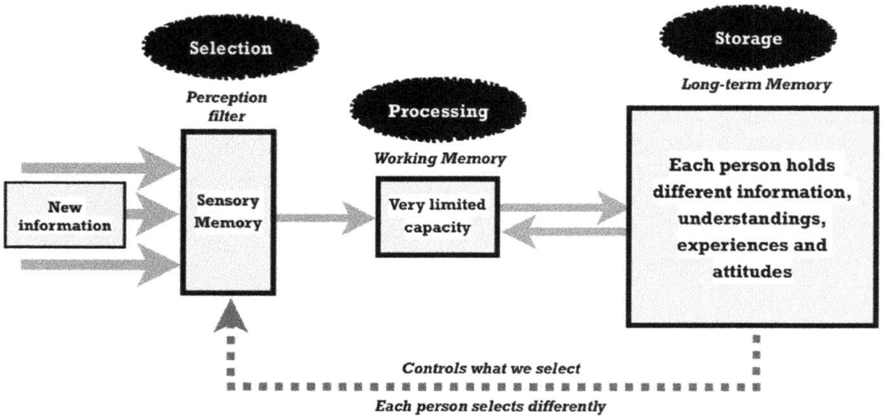


**Figure 2.7** Information flow in the human brain.

Overall, the information flow processes are summarised in Figure 2.7. The model summarises what the evidence has revealed about the way information moves within the human brain. It must be stressed that this model is not speculation. There is extensive evidence from psychology, medicine, neurobiology and education which, together, generates the model. However, the real evidence that the model does represent reality—at least in strong measure—lies in the fact that the model has been used extensively to predict. Therefore, there are many studies which have been undertaken that have asked questions like: '*if we do this, what does the model predict will be outcomes?*' The evidence shows that the model does predict correctly.

> *'Theories'*
> Sadly, education is full of references to educational theories where it seems that they are seen as little more than sensible speculations and we are free to pick the one we like.
> A genuine theory is based on extensive, replicated evidence and offers the best interpretation (at least for the moment) in making sense of the data.
> See: Reid and Ali, 2020, (pp. 11–12).

Overall, the information processing model is extremely well supported. However, as yet, it does not offer insights about the speed of movement although much is now known about how long information can be held in the sensory memory and the working memory. What we do know is that, after working memory has processed one set of information, it can be cleared for the next set. There is a need for research to be undertaken in exploring the speed by which working memory operates as well as the speed when it clears itself for the next task.

Nonetheless, it does appear that the working memory tires. Work Johnstone directed revealed some amazing insights about lecturing in chemistry classes

(Johnstone and Percival, 1976). In any lecture, the human brain (the working memory) tires and needs a short break to refresh itself. There may two or three such breaks in a traditional one-hour lecture and it can be observed that the entire class seems almost to '*switch off*' to gain the needed refreshment. Some lecturers have observed this and found useful ways to employ the short gap, maybe two or three minutes (Cutts, 2006).

## 2.6 The Perception Filter

The perception filter takes in information gathered by our five senses. In learning chemistry, vision and sound will dominate although smell and touch may be important in some areas. Baddeley (1997) has shown that information is held in this part of our brain for some milliseconds. Selected information is converted to electrical impulses and these are passed to the working memory.

Johnstone realised that the role of the perception filter was to reduce the load of information reaching us to manageable proportions. This principle is vitally important in teaching chemistry. We do not know how individual learners in front of us are selecting what is being allowed into their working memories. On the basis of inadequate understandings of the past, they may ignore what is important and they may only select in what fits their preconceived ideas. In the light of the importance of selection, Johnstone developed practical ways to enable the selection process to operate more efficiently, showing that the extent of understandings gained were enhanced.

Based on the findings of Ausubel (1968) long ago, it is now clear that the selection in the perception filter is *driven by* the long-term memory. What we know, what we understand, what we have experienced—all these control the way new information is selected. Johnstone summarised this by noting that, '*The perception filter must be driven by what we already know and understand. Our previous knowledge, biases, prejudices, preferences, likes and dislikes and beliefs must all play a part*' (Johnstone, 1997, p. 262).

The efficiency of the perception filter is also important. Some learners are able to select in what is central for the task in hand, leaving aside other (albeit perhaps important) information which is not of direct relevance for that task. However, when the selection takes in too much information, working memory then is cluttered with irrelevant information and is easily overloaded, leaving little capacity for thought. Understanding is the casualty.

Johnstone and his team researched this in some detail, looking at what is known as field dependence and field independence. This can be seen as the skill of being able to screen out important, but irrelevant, information when tackling some task. The area arose from work by Witkin and Asch (1948) and can be described as *extent of field dependency*. Extent of field dependency has a considerable impact on extent of understanding and also on assessment performance. There is a very large number of studies and a look at the range of references in a few reveals just how much work has been undertaken (Johnstone and Al-Naeme, 1991; Danili and Reid, 2004; Hindal, 2007; Alenezi, 2008; Hindal *et al.*, 2009; Hindal, 2014; Onwumere and Reid, 2014).

Tinajero and Paramo (1998) offer a very useful review of the extent of the research at that time and they draw out the main conclusions from the work.

Interestingly, the concept of field dependence goes back over 70 years, stemming from work published in 1948 (Witkin and Asch, 1948). The original concept was developed by Witkin and his team (Witkin, 1964; Witkin *et al.*, 1967, 1974; Goodenough, 1976; Witkin *et al.*, 1977; Witkin and Goodenough, 1981) in very different circumstances and they developed a robust and valid test for the skill (Witkin *et al.*, 1971). This is known as the Group Embedded Figure Test and a high score indicates strength in being field independent. The field independent person is able to focus in on the essentials for the task in hand, leaving aside other information.

Witkin and Goodenough (1981) suggested a nice analogy. In tuning an old radio set, we scan the wavelengths (with all their noise) until we can tune in on the desired station. The learner who is the field independent person is able to ignore the '*noise*' and focus on the '*signal*'. What is found consistently is that the person who is field independent *always* has a very large advantage in understanding and assessment. Johnstone interpreted this in terms of the field independent person being much better equipped in selecting what enters the working memory, therefore reducing the possibility of working memory overload. Witkin also found clear evidence that extent of field dependency was a generic attribute and people seemed to be remarkably consistent in this characteristic (Witkin *et al.*, 1974; Witkin and Goodenough, 1981).

Johnstone and Al-Naeme (1991) provide one of the earliest studies that looked at the effect of working memory capacity along with extent field dependency on performance in a chemistry examination (Johnstone and Al-Naeme, 1991). They considered several studies and showed that those with an above working memory capacity and who were field independent were performing markedly better than those with a below working memory capacity and who were field dependent. Working in the context of mathematics, the later review by Alenezi (2008) confirmed that performance can be influenced very extensively by working memory capacity and extent of field dependency.

Johnstone illustrated the simple principle of how extent of field dependency influences extent of understanding. He considered two imaginary students in a chemistry class with similar working memory capacities (Figure 2.8). However, one is highly field independent while the other is highly field dependent.

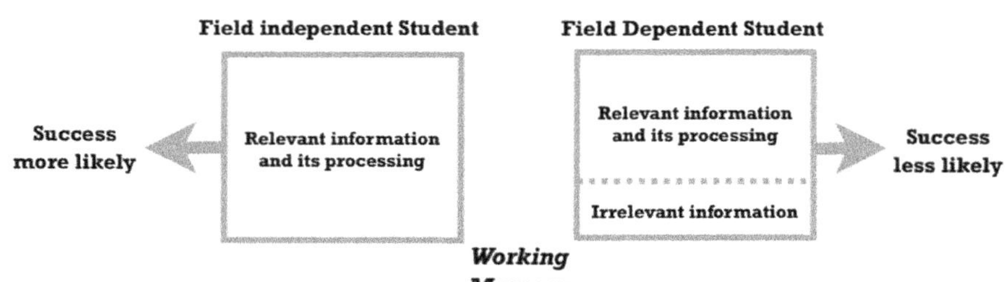


**Figure 2.8** How the extent of field dependence can influence success.

The latter takes into the working memory information what is not directly relevant, cluttering the working memory and hindering the possibility of understanding (or problem-solving, or answering an examination question). The working memory of the field independent student is much less overloaded with extraneous information. This leads to greater success and this happens because the entire working memory space is now available to think about what is required for the task in hand.

There are important lessons for all teaching chemistry: if we do not fill our working memories with unnecessary information, then it is more likely for us to have enough working memory capacity to be successful in the task given to us. Working memory capacity for any individual is fixed. The question is whether field dependency can be enhanced in any way. If that is possible, then understanding chemistry will become easier.

This raises a research agenda for the future. It does seem that early childhood experiences are related to the extent of field dependence. This suggests that there are strategies that can help. However, are these strategies limited to early childhood or are they possible in later formal education? Watching skilled teachers of mathematics is helpful here. Students find topics like algebra-story problems intensely demanding. Skilled mathematics teachers seek to help the learners to focus in on key aspects of the problems, allowing the learners to know *where* to start and the *key issues* that need identified. Does this offer a useful way forward for chemistry?

Onwumere has shown that extent of field dependency develops with age (Onwumere and Reid, 2014). He looked at the age range from 12 to 17 and found that learners were becoming more field independent with age. He undertook this in the context of mathematics where the importance of field dependency seems to be even more important. Previous studies have also found evidence of growth (Witkin *et al.*, 1967, 1974; Gurley, 1984). However, is the growth simply genetic (as with working memory capacity) or can the growth be enhanced? These studies overall strongly suggest the latter.

For the practising chemistry teacher, we need to be aware that different students are operating differently, not only in what they select into their working memories but also the efficiency of that selection. Anything we can do to assist the efficiency of the selection will enhance understanding. One key aspect is to point out the key landmarks in any area where we are teaching. In simple terms, we need to flag up what is important and what is peripheral.

In an interesting study directed by Johnstone, the lecture notes generated by a large number of first year university chemistry students were examined. On average, the students recorded about 10% of what was delivered by the lecturers. However, what they recorded varied enormously, illustrating how selection varied. They found that the quality of the notes was related to the measured extent of field dependency and working memory capacity. Those who could select more efficiently did better (Johnstone and Su, 1994). While lecturing may be efficient in that one teacher can teach a very large of students, this study undermines the assumption that lecturing is very effective in developing understanding. It highlights the need for lecturers to *'flag up'* key ideas and give the learners very clear landmarks in what is being presented so that the learners are able to focus in what is more important and build their notes around these key ideas.

Long ago, Ausubel demonstrated the central importance of previous knowledge and understanding in controlling future understanding. Adapting Ausubel's findings, this was explored in some detail by two researchers, following a first-year chemistry class over six successive years. To help the perception filter to work more efficiently, what were known as *'pre-lectures'* were employed. Pre-lectures are an example of *'pre-learning'*, an idea developed by Johnstone.

**Pre-learning**

Pre-learning involves learners undertaking short tasks *in advance* of the formal teaching. The goal of pre-learning is to make sure that background understandings are in place and that the learners are prepared for what will face them.

Johnstone considered the information processing model. This showed that what was already held in long-term memory controlled the perception filter. By re-visiting what was in the long-term memory, the selection process was predicted to become more efficient, reducing overload possibilities in the working memory.

Johnstone developed pre-labs as well as pre-lectures. The evidence shows how powerful this approach is in enhancing understanding.

Pre-lectures: Sirhan *et al.*, 1999; Sirhan and Reid, 2001

Pre-labs: Johnstone *et al.*, 1994, 1998

For each course, the first lecture was employed in a different way. In an ordinary tiered lecture theatre, students were given a projected set of a few multiple-choice questions that related to key ideas and understandings that they should have brought with them from their school chemistry and which were relevant to the course about to start. They marked themselves.

On the basis of marks gained, the entire group (typically up to 200 students) was invited to divide itself into two groups: those that needed help, those who could offer help. Moving around as necessary, students formed pairs, one able to offer help paired with one needing help. For the rest of the lecture hour, they undertook a series of tasks (projected on screen) working in pairs. These tasks were based on school chemistry, which was relevant to focus on the key ideas that underpinned the course that was about the take place.

It is important to note that the removal of this one lecture reduced the teaching time available for formal lectures by up to 10%. In fact, the study looked at six successive year groups, involving 1043 students overall, an average of 174 each year. In the first two years, pre-lectures were employed. With a new head of class coming in, pre-lectures were discontinued for the next three years. In the final year, a form of paper pre-lecture was offered to students (to be completed, voluntarily, in their own time). Performance in class examinations was followed with six successive classes and the outcomes were quite surprising.

When pre-lectures (or paper-based pre-lectures) were employed, the *least* qualified (school chemistry grades) group of students performed as well as the best qualified. Then there were no pre-lectures. The least qualified performed much worse. Overall, with pre-lectures in place, the evidence suggests that the overall standard of performance increased, despite the teaching time for the lecture courses being reduced by 10%. The findings can be found in two papers (Sirhan *et al.*, 1999; Sirhan and Reid, 2001).

This is another example of the power of the information processing model to predict. The model predicts that, when the perception filter is functioning efficiently, then working memory overload is less likely, leading to better understanding. The pre-lecture activities refreshed the students in the key understandings that they held from past learning as well as marking out central ideas that would underpin the new material that would be given to them. The perception filter would then work more efficiently. Working memory overload would be less likely. Understanding would be enhanced. Examination performance would be improved.

Reid and Ali (2020, pp. 97–101) have listed the key findings from a very large number of studies:

(a) Extent of field dependency can be seen as the ability to perceive the '*message*', ignoring the '*noise*'.
(b) Extent of field dependency can be seen as one aspect of the efficiency of working memory and is controlled by the functioning of the perception filter.
(c) The characteristic can be measured easily using the Group Embedded Figures Test.
(d) Those who are more field independent *always* perform better in academic examinations and tests.
(e) Extent of field dependency grows with age and seems strongly influenced by early childhood experiences.

## 2.7 Key Messages

Much research has led to the development of an information processing model. This model shows the flow of information in the human brain. The perception filter receives information from the senses and selects what to pass on to the working memory. The basis of selection is based heavily on previous knowledge and understandings.

The working memory is central to the entire process. This is the one place in the brain where we think, understand and solve problems. It is very much the working part of the brain, the long-term memory acting as a store, of apparently infinite capacity. The key finding was that of Miller (1956) and it was a surprising finding. The working memory has fixed capacity and the capacity is quite small. The working memory, on average, can hold seven chunks and these chunks may involve information, ideas or processes to be undertaken. What constitutes a chunk is what the individual person sees as one '*unit*' of information.

Individual vary in the capacities of their working memories, with most of the adult (16+) population having between five and nine chunks available to them. The brilliance of Johnstone was being able to see that Miller's findings offered a simple explanation of the difficulties learners have when making sense of a subject like chemistry which is high in conceptual demand at quite early stages of learning.

It is well known that abstract subjects are demanding. However, it is unhelpful to suggest that the reason that conceptual learning is demanding is because it involves the abstract. We need to recognise that:

| |
|---|
| Abstraction is a description, not an explanation. |

Johnstone had now found a simple explanation to account for the difficulties that abstract ideas brought to so many learners. Concepts, by their nature are abstract. Understanding a concept often requires the learner to hold many ideas *at the same time*, to gain any kind of understanding. If the number of ideas to be held *at the same time* is too large, working memories cannot cope and understanding is simply impossible (Chu and Reid, 2012).

Over many years, these ideas were tested by many and the predictive value of the information processing model was demonstrated again and again. Here, we had a simple model that offered explanations, but the model offered ways to explore new ways forward in attempts to help our students make sense of the complex concepts so often seen in chemistry. We shall look at some of the concepts later and show how a considerable amount of chemistry can be understood without invoking so many concepts quite so early.

Johnstone was seeking to address the fundamental question: why is it that understanding chemistry is so often found to be demanding? Two questions followed:

(1) Is there something about the nature of chemistry itself that makes it difficult?
(2) Is there something about how humans gain understanding that makes understanding chemistry difficult?

He now had an answer to the second question but what about the first? This is the theme of the next chapter. This leads to the development of his very simple but powerful triangle model.

## References


Alenezi D., (2008), *A Study of Learning Mathematics Related to some Cognitive Factors and to Attitudes*, PhD thesis, Glasgow: University of Glasgow, http://theses.gla.ac.uk/333/, Accessed 7 November 2019.

Ali A. A. and Reid N., (2012), Understanding Mathematics Some Key Factors, *Eur. J. Educ. Res.*, **1**(3), 283–299.

Ali A. A. and Reid N., (2020), *Making Sense of Learning, A research-based approach*, Berlin: Springer.

Atkinson C., (1983), *Making Sense of Piaget*, London: Routledge and Kegan Paul.

Atkinson R. C. and Shiffrin R. M., (1968), Human memory: A proposed system and its control processes, in Spence, K. W. (ed.), *The Psychology of Learning and Motivation: Advances in Research and Theory*, New York: Academic Press, vol. 2, pp. 89–195.

Ausubel D. P., (1968), *Educational Psychology: A Cognitive View*, New York: Holt, Rinehart and Winston.

Baddeley A. D., (1997), *Human Memory: Theory and Practice*, Hove: Psychology Press Ltd.

Baddeley A. D., (2007), *Working Memory, Thought, and Action*, Oxford: OUP, vol. 45.

Baddeley A. D., (2010), Working Memory, *Curr. Biol.*, **20**(4), R136–R140.

Baddeley A. and Hitch G. J., (1974), Working memory. The psychology of learning and motivation, in Bower, G. A. (ed.), *Recent Advances in Learning and Motivation*, New York: Academic, pp. 47–89.

Baddeley A. D., Eysenck M. W. and Anderson M. C., (2015), *Memory*, 2nd edn, Glasgow: Psychology Press.

Bliss J., (1995), Piaget and after: The case of learning science, *Stud. Sci. Educ.*, **25**, 139–172.

Bloom B. S., (1956), *Taxonomy of Educational Objectives, Handbook I: The Cognitive Domain*, New York: David McKay Co Inc.

Bruner J. S., (1960), *The Process of Education*, Cambridge, Mass.: Harvard University Press.

Chu Y.-C. and Reid N., (2012), Genetics at school level: addressing the difficulties, *Res. Sci. Technol. Educ.*, **31**(1), 1–25.

Cowan N., (2014), Working Memory Underpins Cognitive Development, Learning, and Education, *Educ. Psychol. Rev.*, **26**(2), 197–223.

Cutts Q., (2006), Practical lessons from four years of using an ARS in every lecture of a large class, in Banks, D. A. (ed.), *Audience Response Systems in Higher Education*, Hershey, PA: Information Science Publishing, pp. 65–79.

Danili E. and Reid N., (2004), Some Strategies to improve performance in school chemistry, based on two cognitive factors, *Res. Sci. Technol. Educ.*, **22**(2), 203–226.

Driver R. and Oldham V., (1986), A constructivist approach to curriculum development in science, *Stud. Sci. Educ.*, **13**, 105–122.

Flavell J., (1963), *The Development Psychology of Jean Piaget*, Princeton, N.J.: D. Van Nostrand.
Geelan D. R., (1997), Epistemological Anarchy and the Many Forms of Constructivism, *Sci. Educ.*, **6**, 15–28.
Goodenough D. R., (1976), The role of individual differences in field dependence as a factor in learning and memory, *Psychol. Bull.*, **83**(4), 675–694.
Gurley M. P., (1984), Characteristics of Motivation, Field-independence, Personality Type, Learning Style and Teaching Preference of the Adult Learner as Compared with Traditional Age College Students (Unpublished Doctoral Dissertation, Catholic University of America), *Diss. Abstr. Int.*, **145**, 1268.
Hadden R. A. and Johnstone A. H., (1982), Primary School Pupils' Attitude to Science: The Years of Formation, *Eur. J. Sci. Educ.*, **4**(4), 397–407.
Hadden R. A. and Johnstone A. H., (1983a), Secondary School Pupils' Attitude to Science: The Year of Erosion, *Eur. J. Sci. Educ.*, **5**(3), 309–318.
Hadden R. A. and Johnstone A. H., (1983b), Secondary School Pupils' Attitude to Science: The Year of Decision, *Eur. J. Sci. Educ.*, **5**(4), 429–438.
Halford G. S., Wilson W. J. and Phillips S., (1998), Processing capacity defined by relational complexity: Implications for comparative, developmental, and cognitive psychology, *Behav. Brain Sci.*, **21**, 803–865.
Hindal H., (2007), *Cognitive characteristics of students in middle schools in State of Kuwait, with emphasis on high achievement*, PhD thesis, Glasgow: University of Glasgow, http://theses.gla.ac.uk/5717/, Accessed 9 November 2019.
Hindal H., (2014), *Cognitive Characteristics of High Achievers*, Germany: Lambert Academic Publishing.
Hindal H., Reid N. and Badgaish M., (2009), Working Memory, Performance and Learner Characteristics, *Res. Sci. Technol. Educ.*, **27**(2), 187–204.
Hindal H., Reid N. and Whitehead R., (2013), A Fresh look at High Ability, *Int. J. Instr.*, **6**(1), 59–76.
Hodgson C. and Pyle K., (2010), *A Literature Review of Assessment for Learning in Science*, Slough: NFER.
Hussein F. and Reid N., (2009), Working Memory and Difficulties in School Chemistry, *Res. Sci. Technol. Educ.*, **27**(2), 161–186.
Johnstone A. H., (1991), Why is Science Difficult to Learn? Things are Seldom What They Seem, *J. Comput. Assist. Learn.*, 7, 75–83.
Johnstone A. H., (1993), The Development of Chemistry Teaching - A Changing Response to Changing Demand, *J. Chem. Educ.*, **70**(9), 701–705.
Johnstone A. H., (1997), Chemistry Teaching, Science or Alchemy? *J. Chem. Educ.*, **74**(3), 262–268.
Johnstone A. H., (2000), Teaching of Chemistry – Logical or Psychological? *Chem. Educ.: Res. Pract. Eur.*, **1**(1), 9–15.

Johnstone A. H. and Al-Naeme F. F., (1991), Room for scientific thought? *Int. J. Sci. Educ.*, **13**(2), 187–192.

Johnstone A. H. and Kellett N. C., (1980), Learning Difficulties in School Science - Towards a Working Hypothesis, *Eur. J. Sci. Educ.*, 2(2), 175–181.

Johnstone A. H. and Percival F., (1976), Attention Breaks in Lectures, *Educ. Chem.*, **13**, 49.

Johnstone A. H. and Selepeng D., (2001), A language problem revisited, *Chem. Educ. Res. Pract.*, **2**(1), 19–29.

Johnstone A. H. and Su W. Y., (1994), Lectures - a Learning Experience? *Educ. Chem.*, **31**(3), 75–79.

Johnstone A. H., Sleet R. J. and Vianna J. F., (1994), An information processing model of learning: its application to an undergraduate laboratory course in chemistry, *Stud. Higher Educ.*, **19**(1), 77–88.

Johnstone A. H., Watt A. and Zaman T. U., (1998), The students' attitude and cognition change to a physics laboratory, *Phys. Educ.*, **33**(1), 22–29.

Jung E.-S., (2005), *Attitudes and learning difficulties in middle school science in South Korea*, MSc thesis, Glasgow: University of Glasgow.

Jung E.-S. and Reid N., (2009), Working Memory and Attitudes, *Res. Sci. Technol. Educ.*, **27**(2), 205–224.

Kirschner P. A., Sweller J. and Clark R. E., (2006), Why Minimal Guidance during Instruction Does not Work: An Analysis of the Failure of Constructivist, Discovery, Problem-based, Experiential, and Inquiry-Based Learning, *Educ. Psychol.*, **41**(2), 75–86.

Miller G. A., (1956), The magical number seven, plus or minus two: some limits on our capacity for processing information, *Psychol. Rev.*, **63**, 81–87.

Miller G. A., Galanter E. and Pribram K. H., (1960), *Plans and the Structure of Behavior*, New York: Henry Holt and Co.

*Models of Working Memory: Mechanisms of Active Maintenance and Executive Control*, Miyake, A. and Shah, P. (ed.), (1999), Cambridge: Cambridge University Press.

Onwumere O. and Reid N., (2014), Field Dependency and Performance in Mathematics, *Eur. J. Educ. Res.*, **3**(1), 43–57.

Osborne R. J. and Wittrock M. C., (1983), Learning science: a generative process, *Sci. Educ.*, **67**(4), 489–508.

Ozmen H., (2004), Some Student Misconceptions in Chemistry; a literature review of chemical bonding, *J. Res. Sci. Educ. Technol.*, **13**(2), 147–159.

Pascual-Leone J. and Smith J., (1969), The encoding and decoding of symbols by children: A new experimental paradigm and a neo-Piagetian model, *J. Exp. Child Psychol.*, **8**, 328–355.

Reid N., (2006), Thoughts on Attitude Measurement, *Res. Sci. Technol. Educ.*, **24**(1), 3–27.

Reid N., (2010), A Scientific Approach to the Teaching of Chemistry, *Chem. Action*, **90**, 29–44.

Reid N., (2015), Attitude Research in Science Education, in M. S. Khine (ed.), *Attitude Measurements in Science Education*, 2nd edn, Charlotte, NC: Information Age Publishing, pp. 3–46.
Reid N. and Ali A. A., (2020), *Making Sense of Learning, A Research-based Approach*, Berlin: Springer (in press).
Reid N. and Skryabina E., (2002), Attitudes Towards Physics, *Res. Sci. Technol. Educ.*, **20**(1), 67–81.
Reid N. and Yang M.-J., (2002a), Open-ended Problem Solving in School Chemistry: A Preliminary Investigation, *Int. J. Sci. Educ.*, **24**(12), 1313–1332.
Reid N. and Yang M.-J., (2002b), The Solving of Problems in Chemistry: The More Open-ended Problem, *Res. Sci. Technol. Educ.*, **20**(1), 83–98.
Scerri E., (2010), Comments on the recent defence of constructivism in chemical education, *Chem. Educ. New Zeal.*, 15–18.
Silverman L., (1989), The Visual-Spatial Learner, *Prev. Sch. Fail.*, **34**, 15–20.
Silverman L., (2002), *Upside-Down Brilliance: The Visual-Spatial Learner*, Denver, CO: DeLeon Publishing, *School Dolphin News*, pp. 6–7.
Sirhan G. and Reid N., (2001), Preparing the Mind of the Learner - Part 2, *Univ. Chem. Educ.*, **5**(1), 52–58.
Sirhan G., Gray C., Johnstone A. H. and Reid N., (1999), Preparing the Mind of the Learner, *Univ. Chem. Educ.*, **3**(2), 43–46.
Solomon J., (1994), Rise and fall of constructivism, *Stud. Sci. Educ.*, **23**, 1–19.
SQA, (1998), *Comparability Study of Scottish Qualifications and GCE Advanced Levels, General report and Subject Reports*, Dalkeith: Scottish Qualifications Authority.
St Clair-Thompson H. L. and Botton C., (2009), Working memory and science education: exploring the compatibility of theoretical approaches, *Res. Sci. Technol. Educ.*, **27**(2), 139–150.
St Clair-Thompson H. L., Botton C. and Overton T. I., (2010), Information Processing: a Review of implications of Johnstone's Model for Science Education, *Res. Sci. Technol. Educ.*, **28**(2), 131–148.
Sweller J., (2011), Cognitive load theory, *Psychol. Learn. Motiv.*, **55**, 37–76.
Sweller J., van Merrienboer J. J. G. and Paas F. G. W. C., (1998), Cognitive architecture and instructional design, *Educ. Psychol. Rev.*, **10**, 251–296.
Tinajero C. and Paramo M., (1998), Field dependence-independence, cognitive style and academic achievement: A review of research and theory, *Eur. J. Psychol. Educ.*, **13**(2), 227–251.
Ud-Din M. N., Reid N. and Malik I. A., (2016), *Training Manual on Educational Assessment*, Islamabad: Federal Board of Intermediate and Secondary Education.
Wadsworth B. J., (2004), *Piaget's theory of cognitive and affective development*, Boston, Mass.; London: Longman.

Witkin H. A., (1964), Origins of Cognitive Style, in Sheerer, C. (ed.), *Cognition: Theory, Research and Promise*, New York: Harper and Row, pp. 172–205.

Witkin H. A. and Asch S. E., (1948), Studies in Space Orientation (III): Perception of the Upright in the Absence of Visual Field, *J. Exp. Psychol.*, **38**, 604–614.

Witkin H. A. and Goodenough D. R., (1981), *Cognitive Styles Essence and Origins: Field Dependence and Field Independence Psychological Issues*, New York, International University Press.

Witkin H. A., Goodenough D. R. and Karp S. A., (1967), Stability of cognitive style from childhood to young adulthood, *J. Pers. Soc. Psychol.*, **7**(3, Pt. 1), 291–300.

Witkin H. A., Oltman P., Raskin E. and Karp S., (1971), *Manual for Embedded Figure Test*, Palo Alto, CA: Consulting Psychologists Press.

Witkin H. A., Dyk R. B., Faterson H. F., Goodenough D. R. and Karp S. A., (1974), *Psychological Differentiation: Studies of Development*, New York: Lawrence Erlbaum Associates Inc.

Witkin H. A., Moore C., Goodenough D. R. and Cox P., (1977), Field-dependent and Field independent Cognitive Styles and their Educational Implications, *Rev. Educ. Res.*, **47**, 1–64.

Yang M.-Y., (2000), *Problems Solving in Chemistry at Secondary School Science,* PhD thesis, Glasgow: University of Glasgow. http://theses.gla.ac.uk/2161/, Accessed 8 November 2019.

CHAPTER 3

# *Johnstone's Triangle: Why Chemistry Is Difficult*

## 3.1 Looking Back

In the last chapter, the research that showed the central importance of the working memory was outlined. Johnstone had started by exploring the difficulties learners were experiencing in understanding chemistry. The data started to reveal that the areas of difficulty were associated with high information load. High information load was understood as the number of ideas that the learner had to hold mentally *at the same time* to achieve understanding, solve a problem or answer an assessment question. Much work from psychology and educational psychology revealed the limited capacity of working memory and this offered a very useful explanation of the key factor that controlled extent of understanding, especially in highly conceptual areas of learning.

Building on the evidence from many sources, Johnstone developed an information processing model. This combined many findings and was soon seen as powerfully predictive. The model led to many studies which offered useful insights, and these insights brought great benefit to learners as they tried to understand subjects like chemistry.

The findings were rapidly picked up by university teachers of chemistry (and some other subject areas) but made limited impact at school level. For example, in an invited presentation to about 50 postgraduate scientists nearing the end of their postgraduate training before entering secondary teaching, I presented some of the Johnstone findings related to working memory and information processing. It was all new to the class and they told me afterwards, as a group, that they, '*had learned more from the one-hour presentation than in their entire Educational Studies courses*'. On another occasion, a science teacher trainer in another university (who was a journal editor at that time) told me he had re-written his course completely in the

Advances in Chemistry Education Series No. 6
The Johnstone Triangle: The Key to Understanding Chemistry
By Norman Reid

Published by the Royal Society of Chemistry, www.rsc.org

light of the Johnstone findings related to working memory. He was baffled at the way teacher education had largely ignored Johnstone's work.

## 3.2 Johnstone's Central Question

At the start, Johnstone was trying to find an answer to the question why is understanding chemistry often demanding? We saw that there were two sub-questions:

(1) Does it relate to the nature of chemistry?
(2) Does it relate to the way humans gain understanding?

The previous two chapters have focussed on the second question. We have seen that the limitations of working memory capacity limits hold the key. We needed to address this question first in that it points to the key to the answer to the first question. Therefore, an answer to the first question can now be addressed.

Arising from his research on the central importance of the working memory with its capacity limitations, Johnstone started to seek an answer to the first question. It all started when he was invited to submit a paper to the *School Science Review* as a contribution to a project in England looking at widening access to chemistry at age 16+. He looked at how we ('*experts*') see chemistry and compared it to how young learners see chemistry. He entitled his contribution, '*Macro- and microchemistry*' (Johnstone, 1982).

He noted how chemists can see things at three levels. We see materials in terms of observable properties and behaviour. However, we can easily summarise these observed features in terms of symbols where we represent what is happening. Then, we go one step further in using atoms and molecules (and sub-atomic particles) to offer interpretations and explanations. However, the representations inter-relate with the interpretations.

He appreciated a major problem for young learners. What they observe (the properties and behaviour) often gives no direct information that leads naturally to representations or explanations. In a later paper, he exemplifies this and noted (Johnstone, 1993, p. 704):

> *We can lay before our students some piles of powder, some black (C), brown (Si) and yellow (S) and say, these are all elements each one consisting of atoms of the same kind. There is nothing that appears to the senses that helps to set out the concept of element.*

Indeed, if we also present a pile of turquoise green crystals ($NiCO_3$), there is nothing to see that can show a learner that this pile of powder involves a compound with three elements.

We generate understanding by taking in observations, experience and information and then the working memory tries to make sense of them. The working memory will draw from long-term memory any information and understandings that might be of help. The difficulty in this kind of example is that there is nothing from past experience stored in long-term memory that gives any key to understanding. Simply by their nature, the concepts of chemistry are not easy to relate to previous understandings.

If we look at the history of chemistry, we see the creativity of humans in employing chemistry for social benefit. Discoveries were made by observation and experience. For example, such observations led to the bronze age before the iron age simply because it was easier to extract copper from its ores using carbon-based fires. In the days of Napoleon, the most distinguished guests at a banquet were served their food on aluminium plates (not gold or silver). At that time, aluminium was prized much more highly. The extraction of aluminium was incredibly difficult (given the discoveries then). There was no attempt to write equations or try to explain what was happening in fundamental particle terms. Almost all chemistry was descriptive, and much proved very useful.

Throughout the 19th century, very able minds were involved in gathering vast amounts of information related to the behaviour of materials. Much was descriptive and it was only when enormous amounts of descriptive material were available that patterns and trends started to become apparent. This observation of patterns started the process that led eventually to the development of periodicity: one of the most powerful ideas in understanding the chemical world.

In the 19th century, the molecular nature of matter was only beginning to be understood and, of course, subatomic particles and their behaviour had yet to be discovered. In our teaching of chemistry, we forget just how long it took some of the greatest minds in the history of humanity to make the discoveries that led to the insights that started to emerge in the 20th century.

Johnstone never argued that we should only teach descriptive chemistry. He did argue that more rigorous teaching could and should be built around descriptive chemistry. He never argued for the trivialisation of chemistry: he believed in rigour. He never discussed the philosophy of chemistry as a discipline. He was focussing on learners and their struggles with chemistry. He never argued that young learners should be taught about the three levels or that models were needed to make links between the levels.

He did argue that *simultaneously* working at the three levels he had described was almost guaranteed to generate working memory overload, making understanding almost impossible for almost all learners. We shall give examples later of successful ways forward which minimise working memory overload.

It took centuries for the concept of an element to become clearer and Mendeleev only produced the first recognisable Periodic Table in 1869. This did not stop humankind taking advantage of all kinds of findings from chemistry for thousands of years. The first suggestion that all matter was made up of very small particles was made by the Greek philosopher, Democritus, in 440 BC. He asked the question, '*if I cut this object in half and then each bit in half, can I go on forever?*' He suggested that there would come a time when he would be down to the fundamental particles. They would be '*uncutable*'. The Greek word for '*uncutable*' is '*atomos*'. This gave us our word '*atom*'.

The idea of the atom arose from logical deduction, not experiment. However, the idea of atoms was not really invoked to *explain* any chemistry until around the 18th to 19th centuries, with the work of John Dalton (1766–1844). The power of the atomic and sub-atomic levels to interpret and explain chemistry phenomena really only developed in the late 19th and early 20th centuries. Indeed, looking at the history of humankind overall, chemistry at the levels of representation and explanation occurred very late. This was simply because the observed evidence often does not give us any easy pointers to the atomic level. Think of young learners today. The face the same problem. The world around them does not give them easy pointers to the molecular world or to the sub-atomic world that chemistry teachers take for granted.

## 3.3 Three Thought Levels

Johnstone (1982, p. 378) presented the three levels in this way (Figure 3.1).

Table 3.1 outlines the three levels that Johnstone described (1982, p. 377).

He then notes that experienced chemists can move freely between the three levels, but the learner is unable to handle all three levels *at the same time.* This is simply a function of limited working memory capacity.

In the context of the project that had invited his contribution, Johnstone argued that:

- Most of the population did not need all three levels of thought.
- There was nothing inferior with the macro level.

He gave several examples of extremely high quality and demanding chemistry that never moved beyond the macro. This chemistry was widely appreciated by young learners for it made sense of *their* world and also gave some insights into the way chemistry worked in obtaining evidence that might bring benefit to the wider society. We shall give examples of some ways by which chemistry can be presented, with rigour and demand, but focussing essentially on the macro level only. Johnstone observes that,

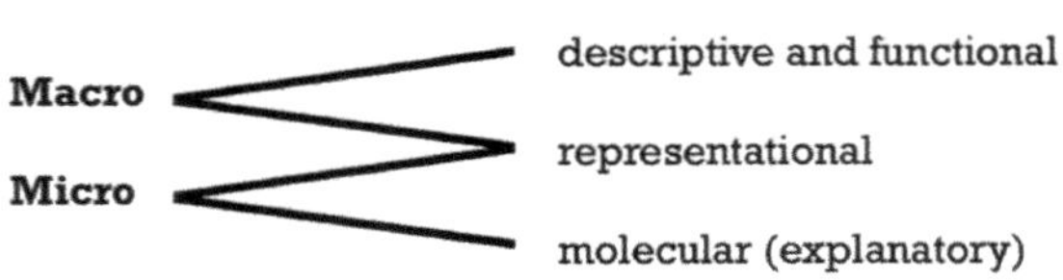


**Figure 3.1** The three levels.

**Table 3.1** How Johnstone saw the three levels originally.

| | |
|---|---|
| Descriptive and functional | *Where we can see and handle materials and describe their properties.* |
| Representational | *Where we try to represent chemical substances by formulae and their change by equations.* |
| Explanatory | *Where we attempt to explain why chemical substances behave the way they do.* |

*'Macro chemistry can provide many opportunities for raising scientific questions, designing experiments to test them, and making useful and valid interpretations'* (Johnstone, 1982, p. 378).

Most of us have undertaken studies in thermodynamics and some of us have taught it. The majority of undergraduates find it hard going. Johnstone points out that thermodynamics moves in the macro and representational and it is not until we move into statistical thermodynamics that the molecular is invoked. It is still a highly rigorous area of chemistry. With a colleague, perhaps one of his greatest publications is a small monograph on thermodynamics where he employed the evidence from the research at that time on how we understand. They then developed the key ideas of thermodynamics at a level suitable for final year school or first year university chemistry students (Johnstone and Webb, 1977). The book is a masterpiece of clear writing. It opens the door to a demanding area of study in an accessible way and makes thermodynamics meaningful in practical ways.

Because we have involved the explanatory level when it is inappropriate, we resort to all kinds of '*explanations*' that do not stand up to any scrutiny. For example, electrons do not move in orbits. Niels Bohr himself showed the impossibility of this. Any atom operating this way might have lifespan of about $10^{-5}$ seconds! Indeed, Johnstone challenged ideas like the '*octet rule*' and the supposed '*desire*' of atoms to achieve this magical octet. The octet is not attained in many entities while the idea of an atom '*wanting*' to lose an election to gain this magical state is undermined totally when learners come to ionisation energies: there is an energy input need to strip out an electron, even from a caesium atom.

It is interesting to see how Johnstone developed much more valid ways to teach atomic and molecular structures in his book of 1981 (Johnstone *et al.*, 1981). It is an interesting but sad observation to note that, after several reprintings over a few years, with the curriculum revised, subsequent books largely reverted to more traditional ways of presenting these topics, ways that are known to generate all kinds of later problems for learners. Tradition and familiarity are difficulty to break. However, this is in no way unique.

In a remarkable monograph published long ago, Professor Jack Linnett (1913–1975) asked the simple questions: why should a negative electron '*want*' to pair up (Linnett, 1964)? What kinds of structures do we obtain if we do not assume this pairing? As an eminent theoretical chemist (a Fellow of the Royal Society), he worked through the logic and mathematics and presented the findings in a very accessible way. Among other things, his model *predicted* the structure of benzene, the paramagnetism of oxygen and the structures of carbon monoxide and nitrogen dioxide. However, after some initial interest, his ideas were just quietly forgotten, despite the overwhelming evidence! Maybe, scientists are not as open-minded as we think?

Johnstone (1982, p. 46) looked at the concept of hybridisation, noting that:

*Statements about hybridisation are often platitudinous mouthing without understanding.* He goes to note that, *It is a strange discipline that takes unreal 'isolated atoms in the gas state' as the norm for discussing electronic configuration and then has to bend the rules by some magic of hybridisation to account for 'reality'. There is nothing in a student's experience and long-term memory to cope with this.*

He goes on to describe much of this as '*pseudo physical reality*' (Johnstone, 1982, p. 46). Some of his '*pseudo realities*' include:

- orbiting electrons
- the octet rule
- '*wanting*' to lose an electron
- hybridisation
- ionic–covalent polarisation.

He asks why we should teach ideas that are unnecessary and need later unlearning? The argument is often suggested that we need these '*pseudo-realities*' at the start before we develop more sophisticated understandings. This is simply not true. Indeed, using such ideas can make it much harder for better understandings to develop later.

It is interesting to note how, with his colleagues, he avoided all these pseudo-realities in his textbook (Johnstone *et al.*, 1981). This was written for students and was their first major incursion into chemistry at the age of around 13–15. None of these pseudo-realities are invoked nor are necessary.

In suggesting these three levels of thought implicit in the nature of chemistry, Johnstone was focussing entirely on the young learner. He was not presenting some philosophical analysis. It was not some '*theory*' of the nature of chemistry. It was simple and practical way to address a question that needed to be faced at that time and still needs faced today:

> On what basis do we select the chemistry to be taught and how it is to be presented?

Over the succeeding decades, considerable research evidence supported and amplified Johnstone's initial thoughts as presented in 1982. Perhaps, his Brasted lecture captures some of the key ideas most clearly (Johnstone, 1997).

His analysis draws out some central issues that must underpin all chemistry education planning, especially at school levels.

(a) Just what chemistry is most meaningful and helpful for the *entire* population of school students?

(b) Given the limitations brought about by working memory capacity, teaching invoking all three levels in unlikely to be successful (in terms of understanding) at early stages.

A third question then has to be addressed.

(c) Should the same chemistry be offered to all school students, irrespective of their likely career directions?

Such questions are part of wider questions that need to be addressed when developing curriculum construction overall. Sadly, it appears that these are rarely addressed. In most curriculum development, teams of subject experts work together to decide what subject content needs covered, usually following the logic of the specific subject in some way. They then determine order and depth of coverage. By contrast, Johnstone was moving thinking in a very different way (Johnstone, 2000a, 2000b).

He was suggesting that we should focus on the learners and their needs in the context of *their* world. This is in no way unique to chemistry. We can consider others subjects as well. One way to look at this is to look at broad subject areas widely taught in schools. The key question is to ask what kind of questions these broad subject areas are seeking to answer.

**Questions addressed by various subject areas**

| | |
|---|---|
| *Sciences* | How do things work and what benefits can arise for humans? |
| *Mathematics* | How do we describe the world around in terms of shape and size? |
| *Creative arts* | What is the impact on humans of wider ways of communicating? |
| *Literature* | What message is literature conveying to us? |
| *Social subjects* | How does human society develop and work? |

In that context, what questions does chemistry address and what are some of the answers it offers? Do these answers have any implications for our lives today? We shall return to these ideas later for there are good examples of chemistry curricula employing this approach and some of them have been discussed (Reid, 1999, 2000).

In his paper of 1999, where he looks further at the nature of chemistry along with the insights arising from information processing research, Johnstone makes an important point (Johnstone, 1999, p. 46):

*Human learning takes place through the senses and our interaction with these sensory stimuli. Concepts are formed through the senses by meeting examples and non-examples and, bit by bit, concepts are refined and strengthened.*

He then goes on to stress that,

*Chemistry concepts are often different from the concepts we meet in most areas of life. They have no tangible aspect: we cannot relate to what our senses tell us.*

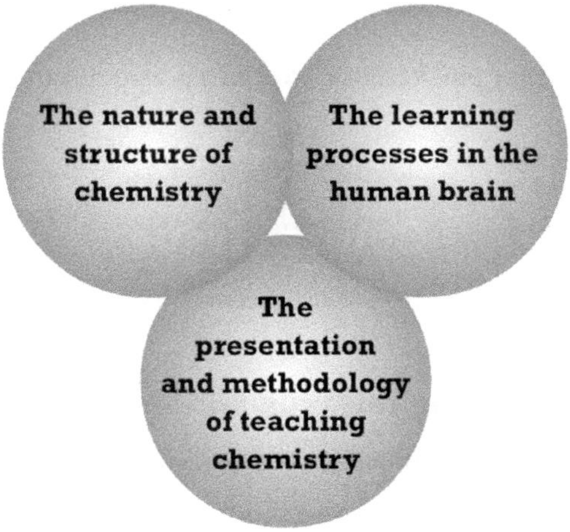


**Figure 3.2** Three interlocking research areas.

He also notes that,

> *We try to make them real by analogy and gross simplification. In so doing, we often lay foundations for misconceptions which in turn need to be unlearned, and the process of unlearning is even harder than learning in the first place* (Johnstone, 1999, p. 46).

Johnstone (1993, p. 705) argues for the place of rigorous chemistry education research to explore three areas in order the help student learning. They can be seen as three interlocking areas of enquiry (Figure 3.2).

In fact, what Johnstone has shown here are the central questions for chemistry education. He has moved our thinking forward enormously with his three insight levels. The work from many sources has given to us an information processing model that has proved to be powerfully predictive as well as stressing that it is the limited capacity of working memory that holds the key to successful understanding.

Johnstone then applied the central idea in developing all kinds of useful strategies that have been demonstrated to enhance understanding while many others have built on his findings. We shall explore many of these practical outworking later in this book (Figure 3.3).

## 3.4 Developing Ideas Further

In a review, Johnstone (1991) pointed out there were three potential sources of the difficulties learners were experiencing when studying a subject like chemistry.

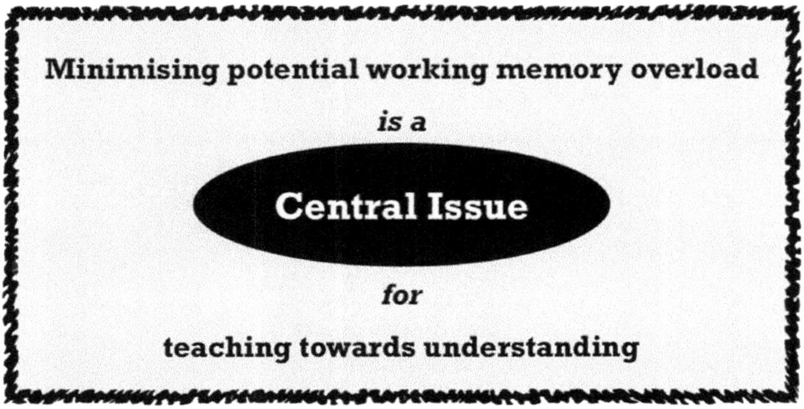


**Figure 3.3** The centrality of working memory.

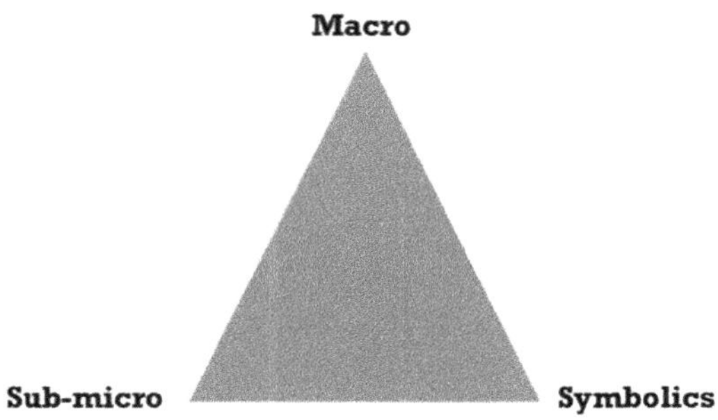


**Figure 3.4** Johnstone's triangle.

(1) The *transmission system:* how we were teaching.
(2) The *learners*: could their brains cope with what was demanded of them?
(3) The *message*: the very nature of the sciences.

Perceptively, he noted that the focus had been too often on the first without adequate recognition of the other two.

In general, learners build up concepts by numerous examples that fit some kind of pattern. The problem with chemistry is that basic concepts like '*elements*' and '*compounds*' and ideas like '*electron, bond energy, photons, structures and molecules*' are not easily open to this kind of approach (Johnstone, 1991, p. 77).

He then introduced the triangle, laid out in its now familiar triangular style (Figure 3.4).

He makes the point that the '*expert*' teacher can work *within* the triangle, drawing in ideas from each corner. The learners cannot do this at the outset—the working memory making it too demanding. He then stresses that the problem arises because we work at all three levels simultaneously with young learners: the important insight relates to the word '*simultaneously*'.

The key here is that the triangle only has the meaning Johnstone intended when we grasp that the working memory and its limited capacity lie at its core. The model was never presented as some kind of philosophical analysis nor as a requirement that we must create bridges between the three levels. The central point is this: with the novice learner, working memory simply cannot cope working at all three levels *at the same time*. Sadly, the central point has often been submerged in some of the literature that followed.

In passing, he suggested three levels for physics and biology (Figure 3.5).

In fact, a much later analysis (Chu and Reid, 2012) suggested that biology might well be a tetrahedron (Figure 3.6).

In this analysis, the *macro* is seen at the organisational level where *molecule* is the equivalent to the *biochemical* in Johnstone's model while *micro* is the *cellular* level. Considerable evidence is brought together to underpin the model.

The biology model was derived from Johnstone's earlier analysis in chemistry and was generated in relation to a study of why genetics was so difficult for young learners to understand. The limited capacity of working memory was shown to be a key factor (correlation with performance gave a value of 0.62 while correlation of performance with extent of field dependency gave a value of 0.48). The point being demonstrated was that the working memory of young learners (in this case aged about 13) simply could not cope working at four levels at the same time. The study went on to develop and test ways to minimise the problems and this showed that the

**Figure 3.5** Physics and biology triangles.

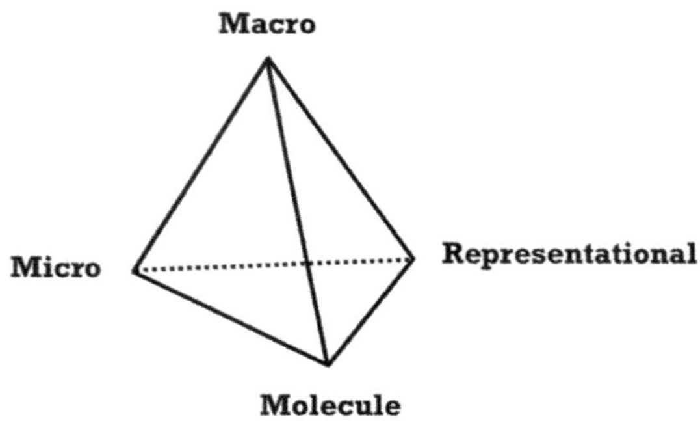


**Figure 3.6** The biology tetrahedron.

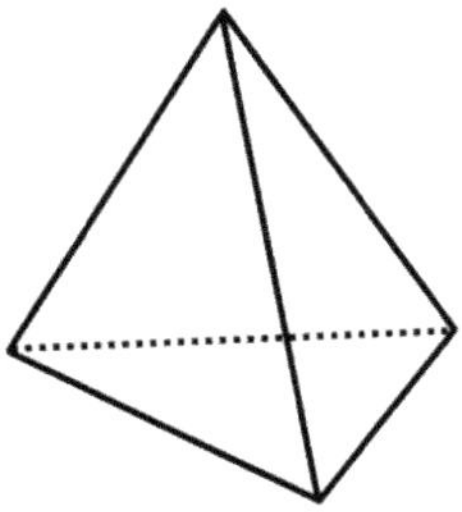


**Figure 3.7** The mathematics tetrahedron.

performance improved quite markedly when possibilities of working memory overload were reduced (Chu and Reid, 2012).

The idea of the physics triangle seems to have had little impact but an analysis in mathematics has generated another tetrahedral model (Alenezi, 2008; Ali and Reid, 2012, p. 284), shown in Figure 3.7.

In the mathematics study undertaken by Ali, a wide range of issues was being considered. Again, the limited capacity working memory was central (correlation of measured working memory capacity with mathematics performance gave values of 0.43 and 0.69 for the two school systems considered in this study). This again illustrates how young learners (aged around 11–13 in this case) simply do not possess the working memory capacity to operate at four levels simultaneously.

**Popularity of Mathematics**

In some countries, mathematics is not popular, and it has sometimes been suggested that showing the applications of mathematics in real life helps.

In fact, evidence shows that this does not work: the working memory cannot handle learning procedures, understanding concepts and seeing applications, while using symbolics, *all at the same time.*

The answer lies in ensuring the procedures and symbolics are '*automated*' before considering adding on understandings and applications (Alenezi, 2008).

Johnstone (1991, p. 82) summarises the key issue when he states:

> *The multilevel thought problem...is capable of rapidly overloading the Working Memory Space with observation, explanation and simultaneous representation...all at the same time.*

This principle applies in any subject discipline but is particularly acute where there are many concepts. Chemistry, physics and mathematics are all concept high. Genetics in biology is extremely concept high and is a well-known difficult area at all levels (Johnstone and Mahmoud, 1980; Bahar *et al.*, 1999; Chandi, 2008; Chandi *et al.*, 2009).

In his paper of 2000, Johnstone summarises much research noting that concepts in chemistry are frequently not open to the senses and not easily related to any ideas already held in long-term memory. Added to that, the three levels of thought easily overwhelm the working memory capacity. He states, '*It is psychological folly to introduce learners to ideas at all three levels simultaneously. Herein lies the origins of many misconceptions. The trained chemist can keep these three in balance, but not the learner*' (Johnstone, 2000b, p. 9). In another paper, he also refers to the three levels of thought, making the same key point (Johnstone, 2000a).

## 3.5 The Triangle in the Literature

Carrying out an internet search using the phrase '*Johnstone's triangle*' will bring up a very large number of '*hits*'. The simple idea has caught the imagination of chemical educators worldwide. In many cases, the central point that Johnstone was making has been grasped. Sadly, in even more cases, the central point has been lost and you can find articles and book chapters on the theme that never mention the working memory and its critical limited capacity. Indeed, some seem to think that learners must be taught the triangle and then shown how to link the levels together. In the process, the central importance and value of the triangle has become submerged in a matrix of other ideas, many of which have limited evidential basis.

Therefore, Johnstone's triangle appears in in many papers, books and internet sites. It has been applauded as a simple way to look at chemistry and it has been employed to justify all kinds of educational developments, from the way chemistry curricula should be constructed across to a justification for various laboratory experiences, from its use as a philosophical analysis of the nature of chemistry as an academic activity across to an emphasis that the model itself should be taught to school chemistry students. The had led some to argue that learners need taught to link the three levels. In this, the role of models is often claimed to be central.

Most studies have looked at the Johnstone triangle and focussed on ways to assist students cope with linking the levels together. With the growing power of new technologies, it is unsurprising that some have turned to the computer for help.

Wu *et al.*, (2001) developed a computer-based visualising tool to help learners look at multiple representations simultaneously and showed how this can assist learners in the development of mental models. Ardac and Akaygun (2004) demonstrated the capabilities of computerised environments to bring together molecular representations and chemistry at the macroscopic level. Tasker and Dalton (2006) also show the usefulness of animations to bring together the Johnstone levels. They overtly built their animations on the evidence arising from information processing, along with cognitive load theory (the psychologists' way of describing working memory overload). In the light of the underpinning principles, great care is needed to bring assistance to learners.

Some recent studies have acknowledged the triangle model but not seen it in terms of its fundamental basis: novice learners find it very demanding to work at three levels *at the same time*, simply because working memory cannot cope. The studies have tended to assume that working at three levels is an essential aspect of chemistry learning and have focussed on ways to link levels. This is unrelated to Johnstone's original ideas and may prove to be an educational cul-de-sac.

Gabel (2005) argues for integrating the three levels. On the basis of extensive research evidence, Johnstone argued that working at all three levels was cognitively far too demanding, and he went on to argue that there was much excellent chemistry which could be undertaken at the macro level. Once that level was secure in the minds of students, the other two levels could be introduced step-by-step, allowing the working memory to cope. Of course, knowledge integration is a desirable goal. How to reach that goal must take the limitations of working memory capacity into account.

Hofgartner (2019) looks at practical ways to link levels but does not see this in terms of the working memory issues. She concludes by noting:

> *Johnstone's triangle has sparked further research and discussion in chemistry education particularly around how it has been interpreted and taught. Regardless, it draws attention to how scientific knowledge is understood at more than one representation or level, enabling students to make explicit links between these levels.*

Nuic and Glazar (2020) refer to the Johnstone triangle and key papers but see the triangle's importance in terms of requiring all learners to be able to work *at all three levels* to have complete understanding of chemistry. They do not seem to see this in terms of problems related to cognitive overload and this is an even more pressing problem at primary school stages.

Keiner and Graulich (2020) focus on the three levels and describe an in-depth analysis of how organic chemistry university student teachers ($N = 9$) explained typical work-up procedures and characterized their activated mechanistic features and transitions between the different representational levels. Their focus is very much on the connections students employ between the macroscopic and sub-microscopic levels in a laboratory setting. They do not relate their findings to working memory considerations.

Gkitzia *et al.* (2020) compared the skills seen with senior school students and third year undergraduates in translating sub-micro understandings into representations. They refer to, '*creating mental images for the corresponding molecular phenomena*' (p. 307). Their focus is entirely on the three levels and the ability to translate across them, not considering the underlying working memory principles. Their paper has usefully collated a large number of references while Scott (2020) has offered a helpful summary of this paper.

**Table 3.2** Developments from the Johnstone triangle.

| Developments | Comment |
|---|---|
| An understanding of all three levels is an essential of education in chemistry at all three levels. | This may be an excellent end goal but working memory limitations make it possible only in stages. |
| The existence of the three levels should be taught to learners. | Why? Is there any evidence this helps? It is more likely to terrify very young learners! |
| The key thing is to enable the learners to link the levels together. | This is a longer-term goal and must be addressed cautiously step by step. |
| The development of models is one keyway to link the levels together. | Models may or may not help much depends on the model working within the capacity of working memory. |
| Modern technologies provide powerful tools to help bring the level together. | Very true, provided that the new technologies take account of limited cognitive resources. |
| There is a need to develop as many '*methods*' as possible to help students to link the levels. | Very true, provided that the '*methods*' take account of limited cognitive resources. |

The number of studies is considerable, and some contain references to many more. However, in many of them, the central principle stated by Johnstone has often become somewhat peripheral. Most publications recognise that the triangle model is a clever insight but go on to make a range of rather different arguments. Some of these are summarised in Table 3.2, with comments alongside.

In all this, we need to remember that the model was developed by Johnstone to bring together much evidence that showed that the central reason why chemistry is difficult is that its highly conceptual nature places very high burdens on limited cognitive resources. In simple terms, our working memories cannot cope when faced with situations which are very high on information. Along with some other areas of study, chemistry is highly conceptual. Understanding concepts requires many ideas to be held *at the same time* in the working memory and this places great burdens on young learners.

Earlier, we saw the central question Johnstone was addressing in his sustained research programmes and how this generated two questions about why learners find chemistry so demanding:

(1) Does it relate to the nature of chemistry?
(2) Does it relate to the way humans gain understanding?

We now have answers:

*Nature of chemistry*: Handling three levels of thought *at the same time* places great burdens on limited working memory capacity.

*Human understanding*: Depends on the fixed and limited capacity of human working memory.

We can see that the answers to the two sub-questions are linked. It all centres on the research related to the central role of working memory in thinking, understanding and solving problems. This central role has sometime become submerged in later writings about the Johnstone triangle while, for some reason, the central role of limited working memory capacity has not become central in teacher education despite the overwhelming evidence related to the insight.

Figure 3.8 shows how the answers to the questions come together.

The sheer number of papers and books (across education, medicine and psychology) consistently confirm the central finding. Indeed, the evidence shows that, in education, it is this finding which Johnstone brought to the world of chemistry education that underpins most of the current issues related to science education today.

Looking at the various reactions in the way the Johnstone triangle has been understood reflects a nice example of how we interpret what comes to our senses, the interpretation being controlled by what we already know and understand. Therefore, the triangle has been seen as a way of looking at chemical literacy, seeing the descriptive at one level, with the qualitative explanation and quantitative explanation following. However, the symbolic level can scarcely be described as qualitative and may, indeed, involve the quantitative.

It has been used as the basis for justifying laboratory work or as the basis for all kinds of ideas related to the use of models in seeking to understand chemistry. Overall, readers of Johnstone's papers have constructed their own understandings, re-interpreting in terms of their own perspectives. In many cases, these do not reflect the intentions of Johnstone in developing the model: constructivism exemplified:

*We don't see things as they are; we see them as we are* (Nin, 1961).

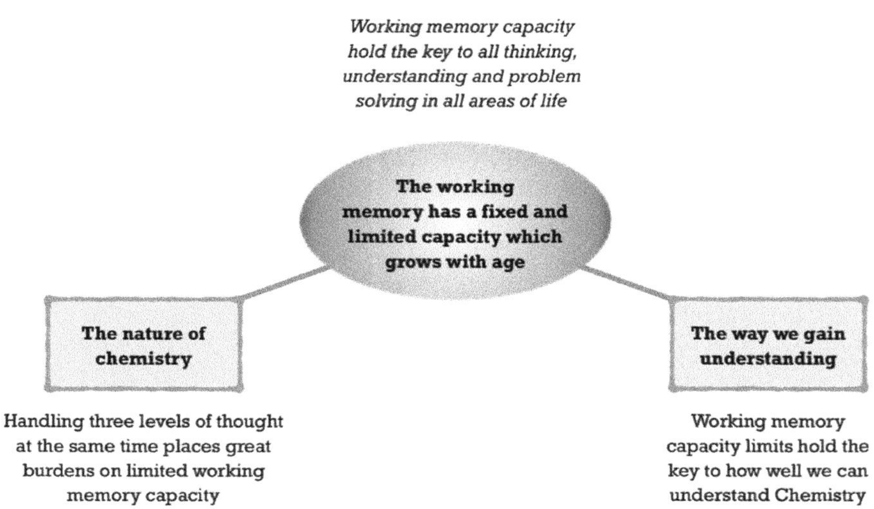


**Figure 3.8** Working memory holds the key.

## 3.6 Specific Insights

Talanquer (2011, p. 179) notes that Johnstone's triangle '*has been one of the most powerful and productive ideas in chemical education for the past 25 years*' but he sees the idea having been expanded, sometimes generating confusions and loss of the central insight. He notes debates over the actual titles at the corners of the triangle. Perhaps the way that captures what Johnstone actually meant is to be seen in Table 3.3.

The actual names are not important. The ideas suggested by Johnstone are important. In various places, Johnstone referred to levels of thought (Johnstone, 1991), components or modes (Johnstone, 1993), forms of subject matter (Johnstone, 2000a, 2000b). Again, the language is not the central issue. The key issue is that the novice learner, with limited cognitive resources, cannot operate at three levels *at the same time.*

**Central Issue**

The novice learner, with limited cognitive resources, cannot operate at three levels *at the same time*.

Some have looked at the triangle idea in a more philosophical way, related to the fundamental nature of chemistry. Therefore, Talanquer (2011, p. 186) refers to '*a complex knowledge space*' while Laing (1999) speaks of four levels: macroscopic, communicative, representational, atomic. However, Johnstone was not presenting some analysis of the nature of chemistry as a subject discipline. He was exploring *why chemistry is difficult*. Johnstone was not arguing over the different ways by which we try to represent reality. He wanted to give direction to help future learners.

Many have adapted the triangle model to start to justify all kinds of discussion about the use of models in chemistry. Inevitably, when we are dealing with highly conceptual subjects like chemistry, mental models start to become important and physical models can sometimes help in the development of helpful mental models. Later, we shall look at some of the brilliant models Johnstone developed, many very early in his career (Johnstone and Morrison, 1864–1969; Johnstone *et al.*, 1981).

Much of the literature related to the place of models in chemistry tends to talk in generalities. By contrast, Bodner and Domin (2000) address a specific issue: the role of *mental* models in the solving of open-ended chemistry problems. They found success was enhanced in terms of the number and

**Table 3.3** The three levels.

| Macro | Representational | Sub-micro |
|---|---|---|
| *That which is directly open to the senses of learners.* | *The ways by which we can try to represent what is thought to be happening at the sub-micro level.* | *That which is not directly open to the senses but interprets what the senses can observe.* |

kind of representations that students could bring to the problem situation. Of particular value is their analysis of what we mean by representation. They note the idea that, '*representations are attempts the brain makes to encode experiences*'. In the context of open-ended problem-solving, they offer a useful insight (Bodner and Domin, 2000, p. 26):

> *An operational definition of an internal representation is that it is the way in which the problem solver stores the internal components of the problem in his or her mind. In contrast to internal representations, 'external representations' are physical manifestations of this information.*

It must be stressed that Bodner and Domin were not considering the triangle model. Their goal was presenting research evidence about how learners solve open-ended chemistry problems. In that context, their consideration of the role of models, mental and physical, is extremely useful. Those students who possessed and could employ mental models more successfully tended to be able to solve such problems better. This relates to, and is consistent with, Yang's earlier work in open-ended problem solving in school chemistry (Reid and Yang, 2002a, 2002b) and the work of Al-Qasmi (2006) in considering open-ended biology problem solving in biology with university students.

Bodner and Domin's work has a reality and precision sometimes lacking elsewhere. They note that external representations can include sequences of words, drawings, lists of information, as well as equations. Equations may be mathematical or chemical in nature. They mention relationships like $PV = nRT$. This can act as a representation, a mental model or, indeed, a powerful chunking device.

Taber looks back over three decades since the Johnstone model was first introduced. He recognises the basis of the model deriving from sound psychological research but acknowledges the idea was been '*reconceptualised in diverse ways*' (Taber, 2013, p. 156). Perhaps, this is a polite way of saying that the model has been employed in ways not consistent with the intentions of the original author! Taber notes that, '*Johnstone (1991) suggested that one reason science was difficult for students, was that it involved what he termed multilevel thought*' (Taber, 2013, p. 156). This captures the fundamental starting point for Johnstone.

Gabel saw the triangle as '*three levels on which chemistry can be taught*'. Gabel suggested that far too much was taught at the symbolic level, tending to encompass both the representational and sub-micro levels. This is almost certainly true, and Johnstone certainly argued strongly for more at the macro level, especially in early stages. However, this not the central point the Johnstone model seeks to make: the novice learner cannot handle all three levels *at the same time*.

Perceptively, Taber notes that the model changed its emphasis in Gilbert and Treagust (Gilbert and Treagust, 2009). Johnstone was describing three levels of thought. This was changed to three ways to conceptualise chemistry situations: a sensory way, a qualitative way, a quantitative way. In the process, the central emphasis of Johnstone was somewhat submerged and was replaced by ideas related to the place of models in the learning of chemistry.

Of course, the place of models in chemistry is important. Indeed, the nature of the kinds of ideas that underpin any understanding of chemistry do not relate easily to the ideas that very young learners have in their long-term memories, based on life experiences. Models are an inevitable consequence of this. Johnstone illustrated that very clearly by showing that there is nothing outwardly obvious in a substance that makes it classify as an element as distinct from a compound. However, the emphasis made by Johnstone was not essentially related to the place of models in chemistry. His central argument might be summarised as:

- Potential working memory overload is the *central* problem that undermines understanding.
- We should not try to work at all three levels *at the same time* with novice learners.
- At the outset, it is better to work at one level until ideas there are secure.
- Much excellent rigorous chemistry can be conducted at the macro (descriptive) level.
- Once the macro level is secure, bring in the representational and then bring in the sub-micro interpretation.
- The representational and the sub-micro can interrelate considerably in that the former can attempt to represent the latter.

Inevitably, models have their place in the development and representation of chemical ideas. These can be physical models, but the ultimate goal is the development of sound *mental* models. This is a key point which is emphasised by Bodner and Domin (2000). When such models are established, the learner is equipped to be able to solve open-ended chemistry problems and apply their understandings to chemistry situations in wider life.

Laboratory work has a major place in developing chemistry at the macro level although laboratory work can also achieve much wider goals (Tsaparlis, 2009). We shall look at the area of models and laboratory work in more detail later, when we consider good examples from the work of Johnstone and others as well as the insights from Tsaparlis.

Summarising and extending the insights from Taber (2013) might lead to a picture (Figure 3.9).

Figure 3.9 offers a model for our thinking. However, we need to remember that it is only a model! Models, physical or mental, may or may not represent reality adequately. What Johnstone was saying was that the novice learner cannot operate in the three boxes in Figure 3.9 *at the same time*: working memory cannot cope.

In his numerous research studies, Johnstone and his colleagues showed the problems associated with language (Cassels and Johnstone, 1977, 1978, 1980; Johnstone and Selepeng, 2001) and the difficulties students had in

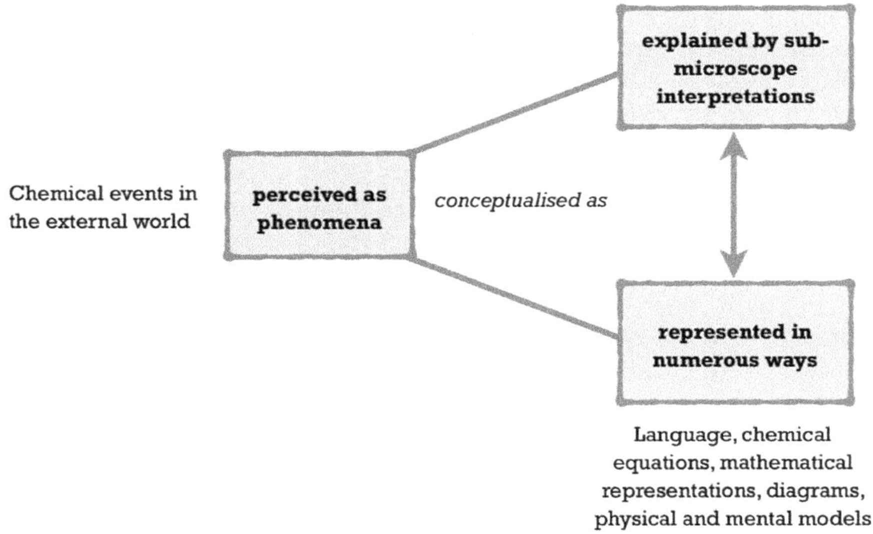


**Figure 3.9** A 'model' of Johnstone's model.

coping with equations as representations (Howe and Johnstone, 1971). Mathematical representations are known to be working memory demanding. Diagrams can help or they can hinder. Yet again, it all depends on working memory considerations (Alenezi, 2008). Johnstone was a master in the development of all kinds of physical models to illustrate chemical ideas and, in his school textbooks, the development of mental models (often through the medium of language and diagrams) is well demonstrated (Johnstone and Morrison, 1964–1968; Johnstone *et al.*, 1981).

However, all these forms of representation have their own working memory demand. On top of that, we tend to employ these various forms of representation to develop sub-microscopic interpretations of chemical events. It is little wonder that the working memory cannot cope!

## 3.7 Models and Representations

Taber illustrates one aspect of the problems of the representational in chemistry stressing that the symbolisms employed need to be understood. There are '*rules*' by which such symbolisms are employed and ways to interpret them. This is like learning a language, parallel to learning the principles of musical notation or even the ground rules of interpreting the meaning of electronic circuitry as presented in a diagram. This all adds to the working memory burden.

The work of Alenezi (2008) in mathematics showed that diagrams can sometimes hinder and sometimes aid understanding. Correctly, she realised it all went back to working memory capacity. If the diagram required too much space in the working memory simply to take it in, then there was little room left for understanding of the meaning conveyed by the diagram. If the diagram, often built up step by step, did not overload working memory capacity, then the diagram aided understanding. The same will be true for chemistry.

Taber also emphasises a feature mentioned by Johnstone. The '*expert*' chemist can move seamlessly between the levels while the '*novice*' often requires all kinds of bridges provided to relate one level another (Taber, 2013). Very often, the sensitive teacher can supply these bridges. Taber is acutely aware of the central importance of the working memory in all chemistry learning. He argues that the Johnstone triangle is important because:

(a) It stresses that there are two levels in chemistry that are needed to describe and interpret chemical phenomena: that is intrinsic to the nature of chemistry (with parallels in the other sciences and mathematics).
(b) Consistent research evidence shows that the human working memory can only cope with a fixed and limited amount of information *at the same time.*

That goes right back to Johnstone's central question: why is understanding chemistry demanding? Is the nature of chemistry or is the way humans come to understanding? It turns out to be both and they are tightly related.

Johnstone never argued that we need to '*water down*' chemistry and, later we shall show how rigorous and demanding chemistry can be taught that still minimises the working memory problem. Indeed, the evidence from physics in Scotland shows that it is possible for learners to be about one year ahead of expectations in what they understand by age 17 and still find their studies fulfilling. We shall tease this out further later as it illustrates another way to approach curriculum design (SQA, 1998).

Taber (2013) takes the same positive and proactive line when he suggests that there are five things that we need to aware of when teaching chemistry. They can be summarised as:

(a) Limit the amount of new information presented *at any one time.*
(b) Help students to relate new information to pre-existing understandings held in long-term memory.
(c) Reinforce and practice ideas and understandings until they become *automated*: the learners are then secure in what they are doing.
(d) Give examples of ways in which chemists have interpreted the macro by means of sub-micro ideas, using appropriate representations.
(e) Support learners as they develop the skills in being able to interpret and describe the macro observations using sub-micro ideas and appropriate representations.

## 3.8 Key Messages

The driving force behind all that Johnstone was doing was his desire to make chemistry accessible and exciting for young learners at all levels. It was his observation that certain topics generated difficulties that led him to direct

**Table 3.4** Implications: the triangle model and working memory limitations.

| |
|---|
| Goals for chemistry education |
| Designing a chemistry curriculum |
| How we present chemistry |
| Importance of language in chemistry |
| The visual and field dependency |
| The place of physical and mental models |
| The role of lab work |
| The role of group work |
| Assessment in chemistry |
| Implications for attitudes and motivation |

many studies to look for ways by which these difficulties might be reduced. He never assumed any specific '*theoretical framework*' nor did he circumscribe himself with research questions that, inevitably, close off options. These approaches hinder genuinely rigorous research enquiry.

By contrast, Johnstone applied the same thought forms that have been so successfully employed for generations in mainstream science research. However, he applied these approaches to the processes of learning. Chemistry research studies the behaviour of matter at the molecular level. Chemistry education research studies the processes in learning chemistry. The same rigour is needed for both areas.

The work led in due time to the understanding that it was the fixed and limited capacity of working memory that controlled extent of understanding and, sometimes, made understanding extremely elusive. Hundreds of studies worldwide have shown that the finding was sound. This then led on to the appreciations that the very nature of a subject like chemistry often required thought at multiple levels and that thinking at three levels was almost always going to create working memory overload, understanding being the inevitable casualty.

The implications of his findings apply to all areas of learning and, specifically, all areas of learning in chemistry. Table 3.4 summarises some key areas.

The following chapters will look at these themes and show how the insights from Johnstone and his triangle model have led to ongoing research programmes and related to ways by which we can move towards the achievement of his original goal: making chemistry accessible and exciting for young learners at all levels. His legacy gives us the agenda for enhancing the chemistry education of future generations.

# References

Al-Qasmi S. A., (2006), *Problem solving in biology at university level*, PhD thesis, Glasgow: University of Glasgow. http://theses.gla.ac.uk/6265/, Accessed 25th November 2019.

Alenezi D., (2008), *A study of learning mathematics related to some cognitive factors and to attitudes*, PhD thesis, Glasgow: University of Glasgow, http://theses.gla.ac.uk/333/, Accessed 7 November 2019.

Ali A. A. and Reid N., (2012), Understanding Mathematics Some Key Factors, *Eur. J. Educ. Res.*, **1**(3), 283–299.

Ardac D. and Akaygun S., (2004), Effectiveness of multimedia-based instruction that emphasizes molecular representations on students' understanding of chemical change, *J. Res. Sci. Teach.*, **41**, 317–337.

Bahar M., Johnstone A. H. and Hansell M. H., (1999), Revisiting Learning Difficulties in Biology, *J. Biol. Educ.*, **33**(2), 84–87.

Bodner G. M. and Domin D. S., (2000), Mental Models: the role of representations in chemistry problem solving, *Univ. Chem. Educ.*, **4**(1), 22–28.

Cassels J. R. T. and Johnstone A. H., (1977), *Language in Chemistry - Research for the Classroom and Beyond*, London: The Chemical Society.

Cassels J. R. T. and Johnstone A. H., (1978), What's in a Word, *New Sci.*, 432.

Cassels J. R. T. and Johnstone A. H., (1980), *Understanding of Non-Technical Words in Science*, London: The Chemical Society.

Chandi S. S., (2008), *Systems thinking as a teaching and learning tool for biology education*, PhD thesis, Glasgow: University of Strathclyde.

Chandi S. S., Reid N., McWilliam A. and Gray D., (2009), Exploring the usefulness of a systems-based model from the perspective of biology students, *Scott. Educ. Stud.*, **41**(1), 68–85.

Chu Y.-C. and Reid N., (2012), Genetics at school level: addressing the difficulties, *Res. Sci. Technol. Educ.*, **31**(1), 1–25.

Gabel D., (2005), Enhancing students' conceptual, understanding of chemistry through integrating the macroscopic, particle and symbolic representations of matter, in N. J. Pienta, M. M. Cooper and T. Greenbowe (ed.), *Chemists' Guide to Effective Teaching*, Upper Saddle River, NJ: Pearson, pp. 77–88.

*Multiple Representations in Chemical Education*, Gilbert, J. K. and Treagust, D. (ed.), (2009), Dordrecht: Springer.

Gkitzia V., Salta K. and Tzougraki C., (2020), Students' competence in translating between different types of chemical representations, *Chem. Educ. Res. Pract.*, **21**(1), 307–330.

Hofgartner R., (2019), Develop deeper understanding with models, *Educ. Chem.*, https://edu.rsc.org/feature/develop-deeper-understanding-with-models/3010519.article, Accessed 2nd February 2020.

Howe T. V. and Johnstone A. H., (1971), *Reason or Memory? - The Learning of Formulae and Equations*, Dundee: National Curriculum Development Centre, Bulletin No. 1.

Johnstone A. H., (1982), Macro and Microchemistry, *Sch. Sci. Rev.*, **6**(227), 377–379.

Johnstone A. H., (1991), Why is Science Difficult to Learn? Things are Seldom What They Seem, *J. Comput. Assist. Learn.*, **7**, 75–83.

Johnstone A. H., (1993), The Development of Chemistry Teaching - A Changing Response to Changing Demand, *J. Chem. Educ.*, **70**(9), 701–705.

Johnstone A. H., (1997), Chemistry Teaching, Science or Alchemy? *J. Chem. Educ.*, **74**(3), 262–268.

Johnstone A. H., (1999), The Nature of Chemistry, *Educ. Chem.*, **36**(2), 45–48.

Johnstone A. H., (2000a), Chemistry Education: where from here? *Univ. Chem. Educ.*, **4**(10), 32–36.

Johnstone A. H., (2000b), Teaching of Chemistry – Logical or Psychological? *Chem. Educ.: Res. Pract. Eur.*, **1**(1), 9–15.

Johnstone A. H. and Mahmoud N. A., (1980), Isolating of Topics of High Perceived Difficulty in School Biology, *J. Biol. Educ.*, **14**(2), 163–166.

Johnstone A. H. and Morrison T. I., (1964–1968), *Chemistry Takes Shape*, vol. 1–5, London: Heinemann Educational.

Johnstone A. H. and Selepeng D., (2001), A language problem revisited, *Chem. Educ. Res. Pract.*, **2**(1), 19–29.

Johnstone A. H., Morrison T. I. and Reid N., (1981), *Chemistry About Us*, London: Heinemann, (reprinted 1982, 1986, 1988).

Johnstone A. H. and Webb G., (1977), *Energy, Chaos and Chemical Change*, London: Heiemann, Also in Italian: *Energia, Caos e Reazioni Chemiche*, (translated by L Cardellini and P Mirone), Piccin, Padova, 2002, ISBN 88-299-1630-7.

Keiner L. and Graulich N., (2020), Transitions between representational levels: characterization of organic chemistry students' mechanistic features when reasoning about laboratory work-up procedures, *Chem. Educ. Res. Pract.*, **21**, 469–482.

Linnett J., (1964), *The Electronic Structure of Molecules*, London: Methuen.

Nin A., (1961), *Seduction of the Minotaur*, https://quoteinvestigator.com/2014/03/09/as-we-are/, Accessed 17 June 2020.

Laing M., (1999), The four-fold jungle, *Spectrum-Pretoria*, **36**, 36–42.

Nuic I. and Glazar S. A., (2020), The Effect of e-Learning Strategy at Primary School Level on Understanding Structure and the States of Matter, *Eurasia J. Math. Sci. Technol. Educ.*, **16**(2), em1823.

Reid N., (1999), Towards an Application-Led Curriculum, *Staff Educ. Dev. Int.*, **3**(1), 71–84.

Reid N., (2000), The Presentation of Chemistry: Logically Driven or Applications Led? *Chem. Educ. Res. Pract.*, **1**(3), 381–392.

Reid N. and Yang M.-J., (2002a), Open-ended Problem Solving in School Chemistry: A Preliminary Investigation, *Int. J. Sci. Educ.*, **24**(12), 1313–1332.

Reid N. and Yang M.-J., (2002b), The Solving of Problems in Chemistry: The More Open-ended Problems, *Res. Sci. Technol. Educ.*, **20**(1), 83–98.

Scott F., (2020), Students' struggle with multiple representations, *Educ. Chem.*, https://edu.rsc.org/education-research/students-struggle-with-multiple-representations/4011073.article, Accessed 2nd February 2020.

SQA, (1998), Comparability Study of Scottish Qualifications and GCE Advanced Levels, *General Report and Subject Reports*, Dalkeith: Scottish Qualifications Authority.

Taber K. S., (2013), Revisiting the chemistry triplet: drawing upon the nature of chemical knowledge and the psychology of learning to inform chemistry education, *Chem. Educ. Res. Pract.*, **14**, 156–168.

Talanquer V., (2011), Macro, submicro, and symbolic: the many faces of the chemistry "triplet", *Int. J. Sci. Educ.*, **33**(2), 179–195.

Tasker R. and Dalton R., (2006), Research into practice: visualisation of the molecular world using animations, *Chem. Educ. Res. Pract.*, **7**, 141–159.

Tsaparlis G., (2009), Learning at the Macro Level: The Role of Practical Work, in Gilbert, J. K. and Treagust, D. (ed.), *Multiple Representations in Chemical Education*, Dordrecht: Springer, pp. 109–136.

Wu H. K., Krajcik J. S. and Soloway E., (2001), Promoting understanding of chemical representations: Students' use of a visualization tool in the classroom, *J. Res. Sci. Teach.*, **38**, 821–842.

CHAPTER 4

# *Johnstone's Triangle and the Curriculum: Making Chemistry Accessible*

## 4.1 Looking Back

In the last chapter, we saw how Johnstone had found that the extent of understanding in chemistry was controlled by the fixed and limited capacity of working memory. Chemistry, by its very nature, often combined multiple levels of thought. He had shown that the teaching of chemistry often involved the macro (that which is directly open to the senses), the sub-micro (that which is not directly open to the senses but interprets what the senses can observe) and the representational (the ways by which we can try to represent what is thought to be happening at the sub-micro level). The limited capacity of working memory makes working at the three levels *at the same time* impossible for novice learners.

If the working memory overloads, then understanding is impossible. The triangle model, with the underpinning understanding of the controlling influence of working memory, therefore offers very precise guidance to enable us to present chemistry in ways that make it more accessible and meaningful for our students. In this chapter, we will look at some areas of chemistry learning and see how the model can point us to more effective ways forward. We start by looking at the place of chemistry in the school curriculum and then consider how Johnstone's triangle gives us direction for curriculum planning, teaching, learning in the laboratory as well pointers to ways forward for assessment.

Advances in Chemistry Education Series No. 6
The Johnstone Triangle: The Key to Understanding Chemistry
By Norman Reid

Published by the Royal Society of Chemistry, www.rsc.org

## 4.2 Why Teach Chemistry?

Before we look at the chemistry curriculum, we need to ask why chemistry is to be taught at school level at all? In some countries, it only made its way into the school curriculum in the late 19th century. However, in many countries, its entry was even later. This illustrates the way school curricula have historically been slow to move into new areas and have retained themes and subjects long past their sell-by date.

In some countries, chemistry as a discrete subject appears from about age 12. In others, it is only introduced a year or two later. In some countries, it is only taught by graduate chemists while, in others, anyone with a science degree (or even a degree with some science in it) is allowed to teach it.[†] In most countries, the chemistry curriculum (like most other subject curricula) is laid down in terms of topics to be covered, following some kind of chemical logic. The implicit message is often that education in chemistry means learning a range of specified chemical facts.

In some documents and papers, it almost seems that the purpose of chemistry education at school level is to generate graduate chemists, along with graduates in other related disciplines that rely on chemistry in some way (like medicine). If we consider a typical secondary school drawing from the entire population range, then it can be shown that the proportion of students leaving to undertake a degree in chemistry is a tiny proportion of that school population. Even if we consider all the related disciplines for which a chemistry qualification at school is essential or useful, the proportion is still very small (less than 10%).

If chemistry is to be taught to generate future undergraduates in key subject disciplines, then the idea of imposing such chemistry at school level on the 90% + who will never use chemistry in this way again is close to inhumane. There must be better reasons for chemistry being taught. Why is chemistry important for all young people of secondary school age (about 12–18)? It comes back to focussing on young people as future citizens, not the discipline of chemistry, nor the generation of future chemists. What is it about chemistry that makes it a valuable area of study for all young people?

Another way of looking at it is to suppose that chemistry was to be eliminated completely from the school curriculum. What problems might this cause? In fact, it is an interesting exercise to look at *any* school subject and ask this question. It is useful way for subject specialists (who by their nature are committed to their subject) to explore the ways in which their subject can make a contribution to the education of future citizens, and all citizens.

For the moment, we are focussing this question on chemistry. What are the unique features of chemistry that make it valuable or, indeed, essential

---

[†]In Scotland, the teaching profession has been controlled for over 50 years by a General Teaching Council that controls teacher training, qualifications, accreditation to teach, *etc.* Strict criteria apply (it is illegal to teach anywhere without registration by the Council). The general pattern is that the Council has contributed to rising standards, along with no significant teacher shortage.

as part of being an educated person today? Of course, possessing qualifications in chemistry can lead to satisfying and useful jobs but that is an inadequate reason for everyone studying chemistry because it will not apply to the majority.

Think of the broader impact of chemistry. Chemical findings have changed lifestyles beyond all recognition. For example, if we removed all synthetic fibres from our clothes it would leave most of us very embarrassed. If we removed organic dyes from our world it would leave our lives physically drab. If we removed key metals from modern artefacts it would have frightening consequences while plastics dominate our packaging, often with dreadful effects on our environment due to human irresponsibility. Analytical chemistry underpins so many areas of modern life, our pharmaceuticals depend on chemistry as do the enormous range of beauty products that we all use. The impact chemistry has on modern societies seems endless.

The sciences seek to enable us to make sense of the world around us: understanding how the physical and biological works: patterns, mechanisms, rationale, hopefully to bring benefit to humans. However, the developments arising from the sciences raise all kinds of important questions. All school students will move on to become citizens in societies that are characterised today by complexity. There are important issues that all the societies of the world face today. All our citizens need to be equipped to take informed decisions. Many of these have a strong science basis. Figure 4.1 illustrates the central goal of education in and through chemistry.

The chemistry taught can be rigorous and will form the basis for anyone moving on to courses in Higher Education. However, re-thinking the curriculum goals in this way allows chemistry to bring benefit to everyone. An example illustrates the general principle:

*In a teaching unit of the past, groups of 15-year-old students had the task of selecting suitable compounds for developing an anaesthetic, the study being*

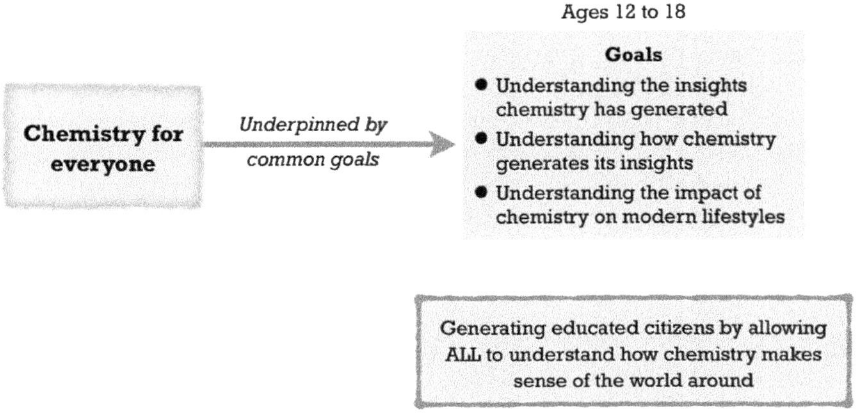


**Figure 4.1** School pathway for chemistry education.

*set in the past. This started by looking at data selected from a fundamental chemistry research study.*

*Carrying out calculations as needed, they had to interpret the data provided in order to identify the chemical substances that were worthy of further testing as potential anaesthetics. The chemistry was rigorous. The implications of the chemistry were life changing.*

*The impact of modern anaesthetics has transformed surgery from the crude butchery of the 19th century to an art-form requiring immense medical skills. However, it all depended on fundamental chemistry research. The students were re-living the experience of turning that research into a positive development.*

Recent headlines have brought the whole area of global warming to the forefront of thought today. It is an interesting, but sad, comment that we are where we are today on this issue because of lack of action that stems right back to the earliest research in the 1970s and 1980s. It might be argued that a better-informed electorate at that time might have pushed their political masters in different directions in the past. Indeed, a scientifically literate political ruling class would help! Perhaps, this is where past science education has failed most significantly? It is a sad that world societies have only been jolted to think about the issue seriously as a result of the committed efforts of a young people today, inspired by a Swedish teenager. It is even more sad that most nations are still not addressing the issue seriously enough to generate positive political action.

The sciences have generated all kinds of amazing developments, but humans have the responsibility to take things forward in ethically responsible ways. There are the implications that arise continually from the genetics revolution. What does this mean for food production or medical advancement?

The science research base needs to be funded. On what basis should funding be determined in the light of other pressing needs? Where are the priority areas? Indeed, can we determine the value of any research in advance given that research means moving into the unknown. We need to remember that the microchip revolution arose as a spin-off from the need for increased processing power, with minimum weight, for space exploration. Could that development, which has transformed modern communications, ever have been predicted in advance?

Citizens with mature insights into the basic science (including chemistry) can assist in making better decisions. Then, there are areas where the management of resources is critical: food and water, accessible electrical energy, world-wide issues of pollution and re-cycling.

**Energy and Chemistry**

In a teaching unit developed more than 40 years ago, groups of students had the task of developing primary energy sources for the UK over a 50-year timespan in 10-year periods (using real data, simplified for accessibility). Surprisingly, the unit was found to work well from age 13 to adulthood.

The chemistry base for most primary energy sources was easily established and the complexity of the decisions was experienced. The students were facing real-life decisions and feedback was given to them telling them of the potential consequences of their decisions.

All school students need to have some understanding of science subjects so that they contribute meaningfully to the ongoing debates that are part of all modern cultures. However, there is a more subtle argument. It is even more important that we develop learners who are able to cope with the kinds of scientific issues of the *future*. Are the present generation of school students equipped to understand the way the sciences work, their goals and their limitations?

Mbajiorgu *et al.* (2017, p. 10) note that there are perhaps three main aspects in learning in the sciences at school level (Figure 4.2).

Let us return again to the Johnstone triangle (Figure 4.3).

If we want future school students to be equipped through their studies in chemistry to contribute to future society in the great scientific issues yet to come, then what chemistry do they need? Perhaps, much needs to be

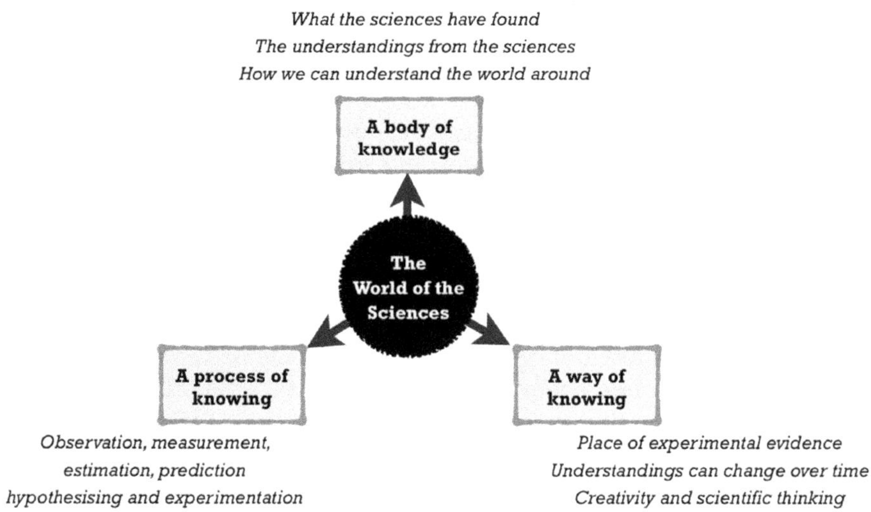


**Figure 4.2** Why study the sciences?

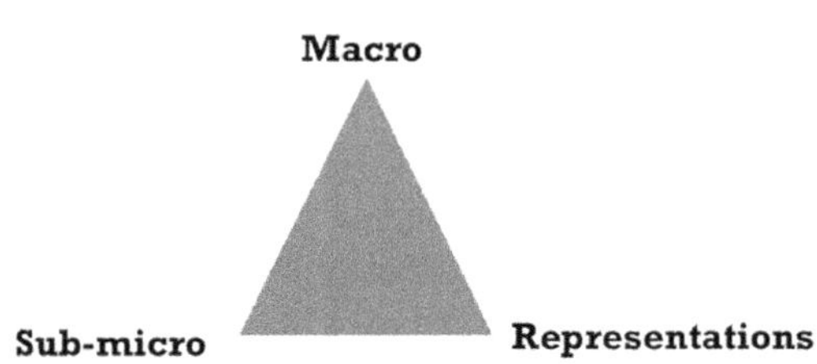


**Figure 4.3** Johnstone's triangle.

dominated by the macro? However, can the macro make sense without moving into the sub-micro, which often requires the use of the representational (symbolics). As part of this, the quantitative may be important.

Johnstone has demonstrated that the limitations of our working memories make it *impossible* for the novice learner to operate at all three levels *at the same time*. He never argued that the three levels were not important although he did stress that large amounts of rigorous chemistry can be based on the macro.

He did argue cogently that we need to take the psychology of the learner into account when deciding what to teach and how to teach it (Johnstone, 2000a). He did argue cogently that we could prune our chemistry curricula enormously (Johnstone, 2000b). This would leave room for much more discussion and debate allowing learners to develop much more mature insights into the nature of the insights that chemistry brings to humankind.

**Overall Goal**
Less information to be memorised by our students and much more thought about the meaning of what is provided.

Our goal must be to generate a population educated in chemistry, not a population who have absorbed half-digested chemical facts.

We can develop the model that was presented by Mbajiorgu *et al.* (2017) a little more (Figure 4.4).

We can summarise this by noting that every school student needs to study chemistry to:

- Understand some of insights that chemistry brings in understanding how our world works.

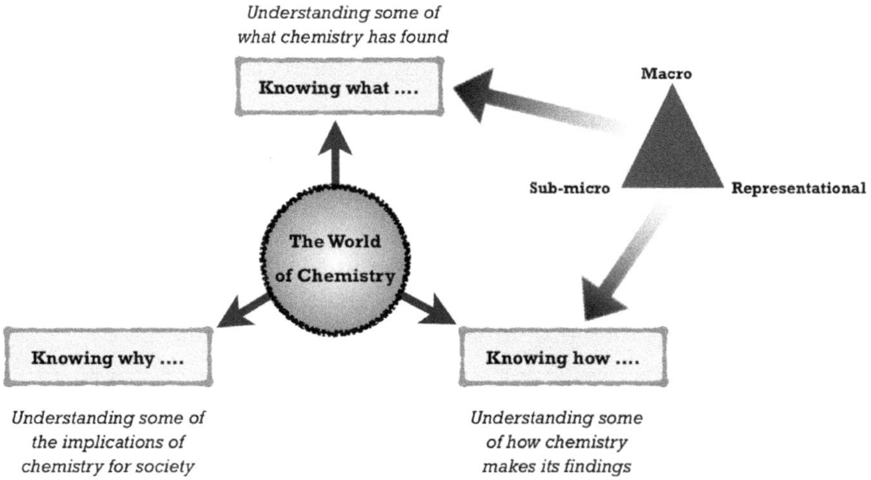


**Figure 4.4** Knowing what, why and how.

- Understand the ways chemistry, as a science, obtains its insights.
- Be equipped, as a future citizen, in the big scientific decisions of the future where chemistry insights play an important role.

## 4.3 What Chemistry and When?

In his later chemistry textbook, Johnstone *et al.* (1981) recognised in the introduction for teachers that the limitations of working memory capacity were critical in making chemistry accessible for young learners (the book was written for 14–16-year-old students). Although he did not use the '*triangle language*', he spoke of employing the minimum of '*atomic theory*' in unfolding the story of chemistry as laid down by a specific syllabus. The story started by a very brief look at atoms and their structure. The goal was simply to establish the idea that links between atoms '*bonds*' were a concentration of negative electrons holding positive nuclei together.

With that established, repulsion of negative charges led to the basic structures that molecules adopt. He employed modelling balloons to illustrate this. This laid the foundation for work much later when electron clouds were involved as a mental model for bonds and lone pairs. For those who progress even further, the approach leads naturally on to quantum ideas and probability. Johnstone avoided incorrect ideas like electrons orbiting nuclei and the specious idea of the octet rule.

With these simple (and highly accessible) ideas in place, carbon chemistry took centre stage. Here, life is easier for we are thinking mainly of three types of atom: carbon, hydrogen and oxygen. Bonds are essentially polar covalent. Ideas like '*sharing electrons*' were avoided. Bonds were simply concentrations of electrons between two adjacent nuclei, therefore holding the nuclei together. Not only did this make chemistry accessible but it also did not require to be '*untaught*' at later stages. We need to ask a serious question:

**Fundamental Question**
Are some of the ideas we invoke in early chemistry completely misleading or even simply incorrect?

Only when these basic ideas were well established (*automated* in the language of Alenezi, 2008 and Shell *et al.*, 2010) are ideas of ionic bonds introduced in the book. The idea that bonding can be presented to students in an artificial dichotomous way of covalent or ionic was not followed.

Polar covalency is the norm in large numbers of molecules.

While polar covalency is the norm in large numbers of molecules, the idea of polarity need not be introduced at this stage. We just focus on the idea that a concentration of negative electrons between atoms holds adjacent positive nuclei together. We make this the central idea and the idea that the electrons might be attracted unequally by the adjacent nuclei emerges naturally.

What this approach achieves is that the development of the ideas of the ionic bond is straightforward while the metallic bond (so often seen as '*weird*' by school students who have been brought up on the covalent–ionic dichotomy) now presents few problems.

**Traditional Approach**

- Bonding as a dichotomy between electron sharing or electric transfer to achieve some mythical octet.
- Young learners have nothing in their experience to make any sense of this. Why an octet? Why do atoms '*want*' it?
- Bond polarity is added later as an afterthought.
- Metallic bonding seems like a strange mystery.

**Outcomes**

- Failure to '*explain*' simple molecules like $SO_2$.
- Delocalisation seen as some strange phenomenon.
- Generates wrong stereo-chemistries.
- Ionisation energies seem to contradict the idea of '*wanting*' octets.

However, it has to be recognised that this was not the first chemistry course the school students had met. There was a two-year course in '*science*' that preceded this chemistry course, which itself lasted two years. This science course involved one teacher teaching all three sciences, irrespective of their qualifications, and the course was developed, *and imposed,* by the then Inspectorate in Scottish schools. It proved to be very unsuccessful. The evidence shows that asking teachers to teach outside their areas of confident knowledge and commitment does a great disservice to learners (Holroyd and Harlen, 1997; Venville *et al.*, 2002; Reid and Skryabina, 2002a).

Many studies showed the rapid deterioration of attitudes towards the sciences during the two year integrated science course (Hadden and Johnstone, 1982, 1983a, 1983b). A later study confirmed this and then demonstrated what it was about a science course (physics) that made it attractive to learners (Reid and Skryabina, 2002a, 2002b). It turned out that there were three factors:

(a) *The quality of the curriculum.*
Does the material to be covered relate to the lifestyle of the students at their age and stage?

(b) *The quality of the teacher.*
Is the teacher a committed expert who possesses empathy for the learners?

(c) *Perceived value.*
Do the learners perceive the value of their studies in terms of potential future careers, occupations and wider life?

The course needs to be perceived by learners to relate to their lifestyle, enabling them to make sense of *their* world. The people best placed to understand how learners see their world, as well their aspirations and problems, are their teachers. Therefore, chemistry teachers must play a key role in the development of chemistry curricula. There are also implications for teacher selection and training. Teachers need to be '*committed experts*'—we need some of the best qualified chemistry graduates to choose to enter teaching and their good qualifications must be rewarded. However, what about the need for teachers who possess high levels of empathy for learners? Is this a selection criterion anywhere in the world today? Can this skill be enhanced? Is it just innate? We know when we have been taught by someone possessing the characteristic. However, can it be objectively discerned in any selection process. Here are the major issues for teacher selection and training.

## 4.4 Ways Forward

In one Scottish school in the early 1980s, the chemistry department rejected the official integrated science curriculum (imposed by the Inspectorate) and built a chemistry curriculum based on the educational research of that time. The course lasted one year and was designed for students about the age of 13. It is an example of a course built almost entirely around the macro level described by Johnstone. Like almost all Scottish Schools, the school was comprehensive[‡] but it drew from an area of some deprivation.

The course (the first formal chemistry course for these students) started by simply looking at how the world was constructed. A Periodic Table had been built in each teaching area, with small medicine bottles showing samples of as many elements as could be obtained (about 50). A beautiful colour photograph of the model Periodic Table can be found in the frontispiece of Johnstone's book (Johnstone and Morrison, 1964).

In the first lesson, the teacher was armed with an English dictionary. The class was asked to estimate how many words were in the dictionary. A typical compact dictionary might contain about 25 000 words. The learners were then asked to think about language, built from 26 letters to form words which were then linked to form sentences. The idea that 26 letters could forms so many words (other languages would multiply this greatly) is macro—it is within their experience.

In a similar way, everything in the world around was made up of about 90 elements (described as different kinds of atoms), linked together in a variety of ways. The goal of the course was then explained: to find what elements were where and to see what humans have done with them to bring useful benefits.

[‡]In Scotland, about 95% attend state comprehensive secondary schools, with fairly equal opportunities for all.

The course then looked at the air, water, sea water and then rocks before a brief excursion into the world of weather. In other words, the course focussed on the physical world around these young people, showing them how to look at the world through the eyes of a chemist.

Rigorous descriptive chemistry followed. Formal chemical language was avoided. However, basic ideas of bond formation and bond breaking were established, laying the foundation for an understanding of the nature of a chemical reaction. Reactivity was observed naturally as students observed that different elements behaved differently. The power of the Periodic Table (whose shape was left unexplained) to summarise patterns of behaviour then became obvious. Observation skills were honed, and the existence of metals and non-metals could be seen. Indeed, much depended on chemical analysis, all at a macro level.

A second Periodic Table had been built in each teaching area. In this, samples of compounds of the elements were displayed. Students immediately asked questions about why there were multiple samples in many transition metal boxes while, almost spontaneously in each class, someone would remark that all the coloured materials were in the middle of the table.

This exemplifies what Johnstone was saying when he argued that an enormous amount of rigorous chemistry could be presented, based almost entirely on the macro level. The amount achieved in one year of study was surprising. However, the course had a somewhat surprising outcome.

The teacher who had developed the course based it on a simple idea: can we enable leaners to make sense for their world from a chemistry perspective? Figure 4.5 outlines the way employed.

We noted that the school drew from a deprived area (former mining and heavy industry had almost all been closed in the early 1980s and unemployment ran at nearer 25%). The course designer realise that the course might be the last chemistry course most of the students would ever undertake. What would be meaningful and useful to them as they moved out into a very uncertain future?

The outcomes observed to arise from this course were indeed utterly surprising. At the end of this one-year macro-chemistry course, it was found

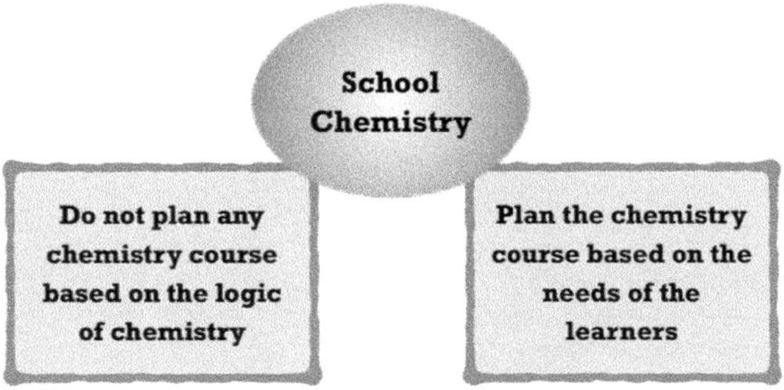


**Figure 4.5** School chemistry.

that almost every student wanted to continue with chemistry. However, the next course was the nationally certificated two-year course (Johnstone *et al.*, 1981). Entry to it was limited to those thought capable of coping. Of the 200 students in the year group, the school increased the numbers allowed to continue on to 60 (a rise of over 50% on previous years).

For those who went on, it was found that the approach had laid great foundations and examination success followed. However, for the 140 who were not permitted to continue, they had learned much of benefit to them and had enjoyed the learning process. Again, all this illustrates the Johnstone argument: the macro can generate highly rigorous and extremely useful chemistry.

This one-year course is an example of what is known as an '*applications-led*' course. This is a course where what is covered, and the order of cover is *determined* by the needs and aspirations of the learners and not by the logic of the subject discipline. There are many examples in existence and two papers have summarised some of the evidence (Reid, 1999, 2000). The area is discussed elsewhere in general terms (Ali and Reid, 2020). An applications-led course in chemistry is not the same as a chemistry-in-context course (a once fashionable idea). In an applications-led course, the content and presentation are *determined* by the needs on the learners. In the latter, the content and presentation are determined by the logic of the subject and, perhaps, *illustrated* by contexts (Figure 4.6).

One of the earliest applications-led courses was a mandatory chemistry course offered as part of pre-nursing training. The teacher found that the would-be nurses neither understood nor enjoyed what he was expected to teach. They simply could not see its relevance to their chosen career pathway. In desperation, he totally re-thought the entire course. He looked at the areas of nursing where chemistry was relevant and then unpacked the

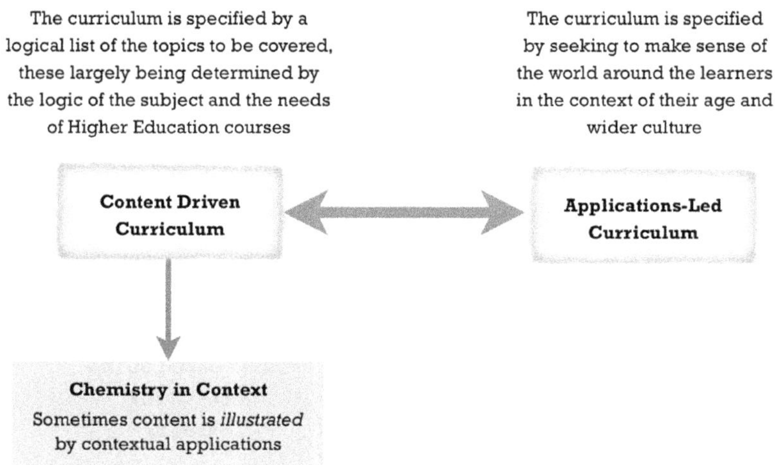


**Figure 4.6** The applications-led course.

chemistry to make sense of what they would be facing in their later training. The same chemistry was covered but now the nurses were motivated, and they understood what was provided for them (Sellinger, 1983).

Earlier, we referred to a physics course in Scotland where physics was so incredibly popular.[§] It was an example of an applications-led course. Eight areas of life were considered, and the physics was then unpacked to make sense of these areas: *e.g.* health, communications, transport, *etc.* It is sometimes asserted that such courses are '*soft options*' and lack rigour. This is simply not true. The evidence from numerous subject areas shows that such courses are usually *more* rigorous than a traditional course, they reach the same destination but with much higher levels of understanding and, as a wonderful bonus, they are *enjoyed* by the learners.

There is quite subtle aspect to any applications-led course. To develop such a course requires careful thought about the learners. This focusses our energies on the learners and the motivation that underpins their learning. It has later been shown that, when learners cannot understand what is presented to them, attitudes deteriorate and motivation sags (Hussein and Reid, 2009; Jung and Reid, 2009). Focussing our thoughts on learners and how they understand is implied in the applications-led paradigm of thought. Johnstone's triangle model has implications in surprising ways. However, we need to remember that:

(a) An applications-led chemistry course is not the same as a teaching chemistry in context.
(b) The same chemistry can be covered in an applications-led chemistry course when compared to traditional course structures.
(c) The chemistry is unpacked bit by bit to make sense of the *world of the learners*.
(d) Evidence shows that such courses are typically *more* rigorous than traditional courses, but they also generate *much more* positive attitudes.

## 4.5 Chemistry's Place in Schools

We are in a position to suggest some ways forward. The ages and stages used in different countries vary but the general structure could be applied anywhere. It has to be stressed that the model shown in Table 4.1 is simply an example and is neither prescriptive nor in any way comprehensive.

There are two major issues that need addressing in chemistry education at school level. However, neither is unique to chemistry and both relate to national policies. Therefore, chemists do not control the outcomes.

The first issue relates to assessment. There is a common view across societies, and widely held by our political masters, that only external

---

[§]Standard Grade Physics, now sadly replaced.

**Table 4.1** Exemplar curricula structure.

| | Age 12–14 | Age 14–16 | Age 16–18 |
|---|---|---|---|
| Year 1 | The elements and where we find them and what we can do with them. | The world of manufacturing and modern materials, anything from clothes, fertilisers, pesticides, anaesthetics antiseptics, pollution, *etc.*, analysis and simple calculations. | Moving into the representational and sub-micro levels in areas of both inorganic and organic chemistry; analysis and more complex calculations. |
| Year 2 | Elements linking, the element of life and energy: carbon chemistry explaining and interpreting; global warming issues. | The world of minerals and rocks: metals, ionic and metallic bonding. | Moving into energy and thermodynamics, kinetics, more organic chemistry, transition metals chemistry, *etc.*, analysis and redox calculations. |
| Johnstone's triangle | Essential macro chemistry. | The macro develops into the simple representational and sub micro. | The representational and sub-micro start to play a large part. |
| Assessment | Entirely in school. | Part in school, part national. | Part in school, part national. |
| Audience | General education, open to all. | Access by student choice, with guidance from school. | |

assessment (controlled by Examination Boards) is rigorous, valid and fair. *The evidence totally contradicts this.* There are areas of the world where there is no external assessment of this nature and where standards are found to be exceptionally high. The overall evidence shows that a combination of teacher assessment and external assessment gives the most valid and reliable outcomes (Gardner *et al.*, 2010; Ud-Din *et al.*, 2016). There are numerous important skills that cannot be measured by externally set paper-and-pen tests. Internal assessment is critical here. The only caveat is that the internal assessment must be set-up with clear measurable criteria. This has been shown to be quite achievable. The worldwide evidence is brought together in Ali and Reid (2020, chapter 14).

The second issue relates to a commonly held view that chemistry (and sometimes other subjects) should be offered in a rigorous way only to those whose career directions are likely to need chemistry and have demonstrated the appropriate ability levels. At its most extreme, chemistry is taught at school to generate future chemists. This argument seems to be implicit in the approach adopted by Gilbert and Treagast (2009) in their discussion of the Johnstone triangle.

We need to consider this argument. First, at school stages, many, maybe even a majority, do not know what their future careers are going to be. Indeed, there is clear evidence, in the Scottish set-up, that students even at university level frequently change (or want to change) their major subject of study during their degree. Johnstone himself started with a view to a degree in physics and changed to chemistry.[¶]

At school levels, young people develop at a wide range of rates. In one extreme case, a boy who needed special needs support for most of his school career suddenly blossomed in his mid-teens. Some years later, he ended up being head-hunted by the US chemistry research establishment following his PhD in Chemistry—an extreme example of late development. However, there are endless examples of young people feeling '*locked into*' a subject when they wished the freedom to pursue something else.

The macro level of chemistry can be highly rigorous. We are in danger of selling chemistry short if we restrict it in any way. Historically, chemistry has made quite major changes to lifestyles. To be an educated person today means that everyone should understand something of the major insights chemistry, as a discipline, has brought to humankind.

Long ago, C.P. Snow addressed the issues of the arts–science divide in education, a major issue at that time in England where he worked (Snow, 1959). He showed the major problems generated by limiting science to *some* in the population and argued that all those leaving school should understand the significance of the second law of thermodynamics. Given that this law determines the direction of change, it is rather fundamental! Johnstone and Webb (1977) showed how the law could be taught and made accessible for all. Indeed, one teacher manual (Reid, 1982) showed how the law could be taught at school level using no more than a bottle of milk. We shall expand on this later in Chapter 7.

## 4.6 Johnstone's Triangle and the Curriculum

The central point that Johnstone was making is that understanding is controlled by the capacity of working memory. In a subject like chemistry, there are three thought levels: the macro, the sub-micro and the representational. The '*novice*' learner cannot handle all three levels *at the same time*. Therefore, it is essential to work at one level only before considering adding on the other levels. He argued that we start with the macro and that there was a large amount of rigorous meaningful chemistry that encompassed that level only.

The starting point must be to look at the way chemistry curricula are constructed (Table 4.2).

[¶]Students do not enter a Department in a Scottish University but are enrolled in a Faculty (or equivalent), a common European pattern.

**Table 4.2** Construction of school chemistry curricula.

| Approximate age range | General Principle |
|---|---|
| Before age 12 | Descriptive only: Piaget showed how the ability to think in abstract terms only developed over the age range of 12–16 (Wadsworth, 1979). Working memory capacity is developing up to age 16 and *controls* extent of understanding. |
| Ages 12–14 | Almost entirely macro although simple equations can be introduced to represent the macro, presented as convenient way to summarise the macro. |
| Ages 14–16 | Build on the macro to develop basic sub-micro insights to interpret the macro, using representations cautiously to summarise insights. |
| Age 16–18 | With the macro now well established (and the basic '*language*' of the sub-micro and representational grasped) the macro can be extended and interpreted using the sub-micro and representational. |
| University chemistry | When new areas of chemistry are taught, it essential that the three levels are not invoked simultaneously but introduced step-by-step. |

It is not going to be the case that we stick to the macro only for some age range and then suddenly, at the next age range, we introduce the sub-micro. The message is that, in areas where learners are '*novices*', we must not overload the working memory and that means we must not invoke the three levels *simultaneously*. It may be that, in one lesson, the macro is the focus while, the next day, the representational is added on.

Organic chemistry is an area where care is needed. It is possible to undertake enormous amounts of rigorous chemistry simply by thinking in terms of the atoms of carbon, hydrogen and oxygen. However, at university level, a mechanistic approach is important. If the macro is not well established already and some key fundamental ideas almost completely automated in the minds of the learners, then the working memory simply cannot cope with the mechanistic level being added on. One study looked at the key underpinning ideas and showed how they were central in enabling first year undergraduates to make sense of their first course in organic chemistry, presented from a mechanistic perspective (Hassan *et al.*, 2004). Here again, the ideas underpinning Johnstone's triangle can be seen to offer the key guidance. This research will be discussed in more detail in Chapter 7.

An example illustrates what can happen at school. This example is described in the textbook used at the time. In a specific curriculum for students of chemistry aged about 14–15, conductivity was being employed as a way to make sense of what was happening during neutralisations (Johnstone and Morrison, 1966, pp. 50–51). The goal of the set of simple experiments was not to carry out rigorous conductivity measurements and then plot precise graphs. The aim was to look at the conductivity *trends* in a solution during neutralisation. The students undertook three experiments. In each case, a very dilute solution of sulfuric acid was placed in a beaker and the conductivity followed as neutralisation took place.

Experiment 1 Adding some copper(II) oxide.
Experiment 2 Adding some copper(II) carbonate.
Experiment 3 Adding some sodium hydroxide solution dropwise from a burette.

On scrap paper, the students then had to draw rough graphs of what they observed. Points were not plotted. All that was needed was the trend to be seen. Figure 4.7 shows the typical graphs obtained.

The working memories of the students were almost completely absorbed in carrying the experiments in pairs, recording what they saw and drawing the rough trends. There was little mental capacity left for thought.

At that point, the teacher brought all the students around the front demonstration desk, bringing their rough graphs with them. The three experiments were repeated very quickly using a demonstration meter which allowed the entire class to see the trends and the teacher drew the graph shapes on the board. There was dialogue and discussion throughout where the teacher drew out from the student group what the trends each meant. The fall of conductivity in all three experiments showed the formation of water, with the loss of hydrogen ions. In the first two cases, once all the acid had been neutralised, adding more copper(II) oxide or carbonate made no further difference for neither dissolved to generate more ions in solution (that was visible: undissolved solid could be seen).

The students could easily grasp the simple ideas that ions could '*carry*' the current but different ions might move at different speeds. This was seen with

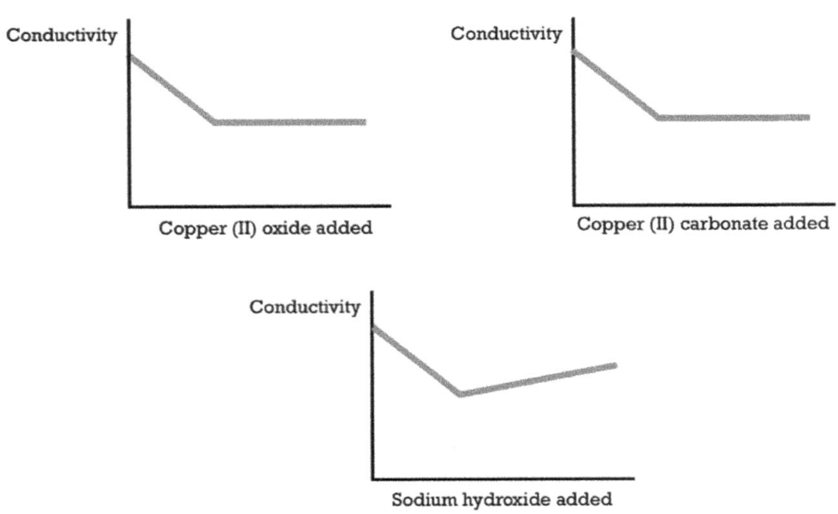


**Figure 4.7** Conductivity trends in neutralisation.

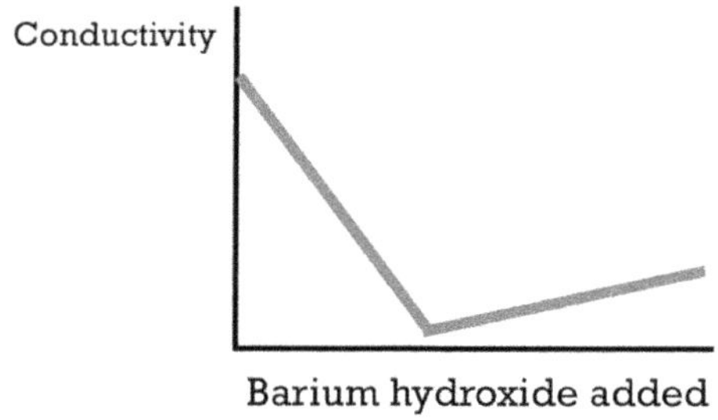


**Figure 4.8** Conductivity trend with barium hydroxide.

the sodium hydroxide where the conductivity did rise again after neutralisation was complete but not as fast as the initial fall. The added hydroxide ion did not move as fast as the hydrogen ion. The idea of '*spectator ions*' (the sulfate ions) was readily apparent.

At that point, the teacher introduced a further experiment. Barium hydroxide was added step by step to the very dilute sulfuric acid. This time the graph fell as before but fell almost to zero before starting to rise when neutralisation was completed (Figure 4.8). A vigorous discussion then ensued as the class tried to work out was happening in terms of the ions in the solution.

The aim was not to teach conductivity nor how to measure it. The aim was to use conductivity evidence as a probe to throw light on might be happening at the ionic level during the various neutralisations. The lesson also illustrated how practical classes may involve both student-conducted work *and* demonstration. The sterile arguments between the relative values of these approaches is pointless: it depends on the aims in mind.

This was one of many experiments that Johnstone developed. The key thing to note here is that it reduces the possibility of working memory overload. The student conducted work focussing on the macro. Once that was secure and was summarised in simple graphs showing trends, the teacher could then build on it to consider the situation at the ionic level.

The approach also opens the door for considerable class discussion as the learners try to make sense of what they have observed (the macro) in the barium hydroxide experiment. This not only encourages careful observation (barium sulfate is insoluble, and cloudiness can be seen but it encourages them to apply the ideas to allow conductivity observations to generate insights into what is happening at the ionic level (sub-micro). The teacher can then draw this together with relevant equations (the representational level). It is a nice example of the Johnstone triangle in operation.

## 4.7 Johnstone's Triangle and Laboratory Teaching

Undertaking laboratory work is an integral part of most courses at all levels in chemistry education. Tsaparlis (2009, p. 109) observes that, '*Contact with concrete examples of substances, their reactions and other properties, through the laboratory and other practical activities, is an integral part of chemical education.*' He argues that laboratory learning is a way to provide bridges between the three levels in Johnstone's triangle. He appreciates that student problems in the laboratory are often affected by working memory overload and he discusses ways to help.

In all this, Tsaparlis has highlighted the potential of laboratory work to build bridges between the macro (what is open to our senses) and the sub-micro and representational. This raises key questions about our aims for laboratory work in chemistry, at both school and university.

Laboratory work started to appear in university courses in the late 19th century but it focussed mainly on the students gaining the practical and technical skills needed by industrial and research laboratories (Blick, 1955; Phillips, 1981, 1983). In schools, laboratory work only started to appear in *some* countries doing the 20th century. This tended to reflect what was happening in universities and there was emphasis on practical hands-on skills.

Overall, with educational opportunities widening greatly during the 20th century, the previous goals were now inappropriate. Indeed, there has often been a lack of clarity about the goals for laboratory work at both school and university levels. Developing practical skills is not central at school level in that it is estimated that only around 1–2% for each science discipline will ever need these skills again later in life.

Numerous sets of goals exist (Johnstone and Wham, 1980, 1982; Carnduff and Reid, 2003; Hanif *et al.*, 2008). One review has looked at schoolwork (Hofstein and Lunetta, 1982) while another review looked at the university level (Reid and Shah, 2007). Bringing the evidence together, it is possible to suggest a set of five goals for chemistry which might apply across all ages (Figure 4.9).

There are some important observations that such a list of five goals can generate:

*Making the chemistry real.* It is too easy for chemistry to be reduced to a strange language of symbols and diagrams, largely meaningless to learners

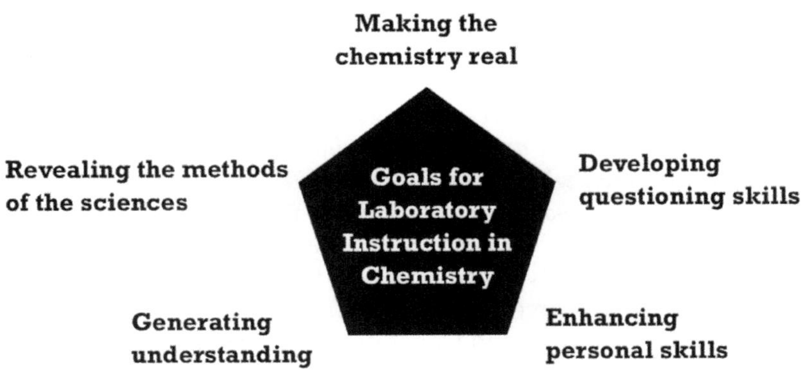


**Figure 4.9** Goals for school laboratory work.

and unrelated to wider life. This pattern is very evident in many developing countries where the laboratory provision is inadequate (Reid and Shah, 2010). The laboratory is where chemistry is to be seen. However, even here, it is too easy for learners to see the laboratory experience in artificial terms. Real chemistry can be seen in every area of life, from the cooking of food across to the working of the internal combustion engine, from the growing of a garden plant across to industrial production. Laboratory work, teaching *and wider life* must be tightly integrated together. Of course, the laboratory is artificial in that it simplifies life processes by employing materials of specified purity.

*The methods of the sciences.* One of the great contributions of all the sciences is to provide a way by which humans interrogate the world around them. That is the way of experimentation, hypothesis formation and testing. Understanding is developed from experimental evidence. This is not to suggest that this is the only evidence that is admissible or that experimental evidence is the best in all situations. However, experimental evidence is a robust way to gain understanding. Therefore, experiments should never be used to *illustrate* theories—that is not the way the sciences work. In fact, using laboratory work this way undermines insights on the experimental as the *key source* of evidence that generates our understandings. The way the laboratory is planned must allow the learners opportunities to make observations and measurements and then seek to interpret them.

*Questioning skills.* Opportunities are needed for students to ask questions, to discuss and argue. There need to be times when the teacher poses questions and then leaves the students to work these through. Ideas and understandings may need challenged while alternative mental models to interpret data need explored. In experimentation, we are asking questions of the world around and, from the way that world behaves in our experiments, we are seeking to develop understanding of the mechanisms underlying the molecular world. Questioning is central.

*Understanding.* Often, learners are so mentally involved in '*doing the experiment*' that there is no mental capacity left to gain insights and understandings. In this, the current fashion of writing laboratory manuals has to be rejected. A series of papers outlines the way evidence has built up (Johnstone and Wham, 1979, 1982; Johnstone and Letton, 1988, 1990, 1991). Johnstone and Al-Shuaili (2001, p. 49) observe that, '*Worksheets and blow-by-blow manuals are still alive and healthy, leading to apparently efficient coverage of laboratory activities, while missing much of the point of what undergraduate laboratories have the potential to achieve*'. Following '*recipes*' generates students trained to follow recipes! The laboratory is a wonderful place where understandings can be developed. It is here that Johnstone's triangle model is so helpful. We need careful preparation *before* the laboratory to minimise working memory overload in the laboratory. We need careful interpretation of what has been done, with the discussions and debate about its meaning, often *after* the laboratory. Report writing usually undermines this: here, we often focus on '*right*' format and the gaining of '*right*' answers. This is very unhelpful.

*Personal skills.* We often forget that, at school level, we are seeking to develop individuals for their future beyond education. Skills like team-working, planning, time management, discussion and debate can be developed in a laboratory situation. They will *only be developed* if the students are working in teams (or pairs), are allowed to plan what they are going to do, are given the freedom to manage their time and enough time to engage in genuine argument, debate and discussion.

In passing, the assessment of laboratory work often completely undermines some of the goals. A report compiled after the laboratory class often gives credit for the wrong things when considering the goals for laboratory work. Observation of the conduct of laboratory skills is also irrelevant at school level. How many of these laboratory skills will be relevant throughout life? Assessment so often distorts the enormous value of laboratory work. Looking at all the evidence, there are some simple practical pointers that can help us to ensure that the laboratory experiences of learners are more productive.

- There is a need to re-think laboratory manuals (mainly university) or prescribed laboratory experiments (mainly school).
- There is a need to re-establish agreed purposes for all laboratory work.
- Laboratory assessments must be re-designed to reflect the agreed goals.
- The limitations of working memory capacity must be central in all planning if the experience is going to be able to enhance understanding. In this, the triangle model must be considered.

Early in his career, Johnstone became aware that university laboratory experiences were not generating relevant cost-effective outcomes for the learners. In fact, little learning was taking place at all (Johnstone and Wham, 1979). In simple terms, most of the mental capacity was being employed simply in carrying out the experiment and then writing it up. This left no mental capacity for understanding. He described a laboratory as a '*noisy*' place, with the students bombarded with information from many sources. Overload was almost inevitable (Johnstone and Wham, 1982). In fact, the working memory problem applies in all learning where hands-on activities are involved but understanding is sought.

In the light of the evidence, Johnstone developed the idea of *pre-learning*. He applied this to laboratory situations (Johnstone *et al.*, 1994, 1998) as well as lectures (Sirhan *et al.*, 1999; Sirhan and Reid, 2001). The key is to prepare the students *before* they entered the laboratory. Pre-learning involves learners undertaking short tasks in advance of the formal teaching. The aim is to ensure that background understandings are in place and that the learners are prepared for what will face them. The aim of pre-learning is to seek to reduce the information load that the working memory has to handle *at the same time.*

In a pre-laboratory task, it is possible to revise background ideas, to introduce unfamiliar apparatus, to direct learners to what are the key things to observe carefully or the possible outcomes that might occur that could generate new insights. When the idea was tested in university laboratory courses, the power of pre-learning to enhance understanding was shown to be quite remarkable Indeed, open-ended projects were then able to be introduced into laboratory work, enriching the laboratory experience powerfully (Johnstone *et al.*, 1994). Attitudes towards laboratory work were also enhanced (Johnstone *et al.*, 1998).

An example illustrates a useful way forward:

**Example**

Students worked in a group of four (sharing the workload) to develop a calibration graph relating to the formation of the phospho-molybdate complex (determined colorimetrically) with concentration of phosphate ion in standard solutions.

They then estimated the concentration of phosphate ions in solutions taken from various sampling points in a local river.

As a group, working with a map, they tried to relate the findings to pinpoint the sources of the pollution.

Pre-laboratory experiences are now widely found in much university laboratory work and ways to develop them have been outlined (Carnduff and Reid, 2003). Overall, time is saved in that the student time to complete the pre-laboratory generates an even greater saving in time *within* the laboratory. Pre-laboratories of this form probably have limited application at school level but the *principles* underpinning the pre-laboratory idea are highly relevant.

Shah developed the idea in a teacher-training context in a country where laboratory facilities and expertise in using the laboratory by teachers are often absent. The goal of the study was not to replace laboratory work but to enable future teachers to use laboratory learning much more effectively (Reid and Shah, 2010).

Shah *et al.* (2007, p. 76) make an important point:

> *Laboratory work in chemistry is an expensive activity. Laboratories are costly to build, to equip; academic and technical staffing, instruments and consumables are a drain of resources. The perception is that it is becoming increasingly difficult to provide students with a high-quality conventional laboratory experience. It is perhaps time to think about the laboratory in terms of quality of experience rather on assessment of skills.*

This stresses the importance of re-thinking our goals carefully. It also alerts us to the real possibilities in that laboratory experiences are often ineffective as learning situations simply because we are not allowing for the limitations for working memory capacity. In a university setting, Carnduff and Reid (2003, p. 31) observe:

> *To change the experience, you don't need to change the experiment, just what you do with it.*

Perhaps, this is a lesson for school laboratory work as well.

## 4.8 Johnstone's Triangle and Teaching

In most countries, the way the school chemistry curriculum is developed needs major overhaul. The normal pattern is that those qualified in chemistry, often academics, work together to develop school curricula, bring content up to date and stress the key landmarks of the subject. It is rare for practising teachers to be involved and even rarer for them to have any *influential* involvement.

Inevitably, those who develop the school curriculum are committed to chemistry. As a result, they have great difficulty in understand the *majority* of the school population who may neither understand nor appreciate that commitment. The outcomes are curricula that reflect the logic of the subject discipline and, very often, not the needs of the young learners. Teachers are then expected to follow a curriculum (and often visiting inspectors make sure they do!) which is, in varying degrees, inappropriate for many.

School education is education. The goal is to equip future generations to develop their full potential, and to be able to be fully participating members of their society. The curriculum is there to offer a framework to bring maximum benefit to all the learners. It is not there to force the learners into the often-artificial logic of a subject discipline as seen by experts in that discipline (Figure 4.10).

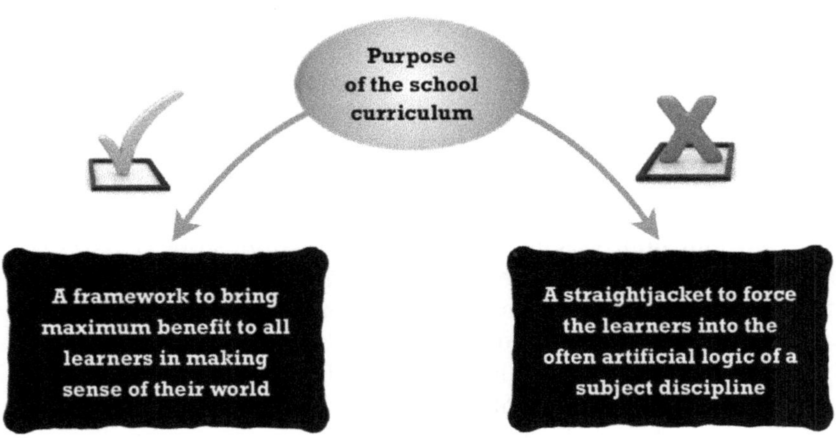


**Figure 4.10** Purpose of the school chemistry curriculum.

The brilliance of Johnstone was that he never lost sight of the learner. Education is the servant of the learner and chemistry education is no different. Nonetheless, Johnstone also never lost sight of the academic rigour and immense human achievement as seen in the progress of chemistry over the centuries. It is beautiful example of a cultural enterprise that has brought enormous benefit to wider humanity and has combined rigour with enormous commitment and sacrifice on the part of many of the great leaders of the past and today. In simple terms, Johnstone believed in academic rigour.

The fundamental issue today is what the chemistry teacher can do when faced with curriculum and assessment structures that are often insensitive to the real needs of young learners. It is here that chemistry teachers need to develop a deep sensitivity for the young learners before them and to be *subversive* in the way they adapt the curriculum straight-jackets in order to give these learners the best experiences possible.

The remarkable teaching changes that enabled nurses to make sense of chemistry (Sellinger, 1983) and the way the curriculum for 13-year-olds was totally re-cast both show what is possible (Reid, 2000). In another country, the textbook dominated all teaching and learning. By re-casting sections of the textbook to allow for working memory limitations, the study showed a quite remarkable improvement in levels of understanding with a concomitant generation of more positive attitudes among learners towards the study of chemistry (Hussein and Reid, 2009). This was achieved simply by re-designing written materials to minimise working memory overload. There was no change to the content covered, the time allocated, the assessment used, and no training was offered to teachers. Perhaps, this points to the need to re-think the way textbooks are structured and to re-establish the central place of working memory in the development of ideas.

In his textbook of 1981, Johnstone stated clearly in the introduction addressed to teachers that the possibility of working memory overload was very real. The way the textbook was structured was an attempt to minimise that overload and make the chemistry more accessible for the learners (Johnstone *et al.*, 1981). This principle has been applied in several studies and always shows the same outcomes: increased understanding (Danili and Reid, 2004; Chandi *et al.*, 2009; Hussein and Reid, 2009; Chu and Reid, 2012).

The way Johnstone was involved in a textbook (Johnstone *et al.*, 1981) that showed ways by which memory overload might be minimised points to ways forward. The goal is to educate *through chemistry* and the evidence shows that, when we do this, we also succeed in educating *in chemistry* to achieve high standards of understanding. This reflects a central principle that underpinned Johnstone's thinking (Figure 4.11).

Nonetheless, there is another problem area that we need to face: that of assessment.

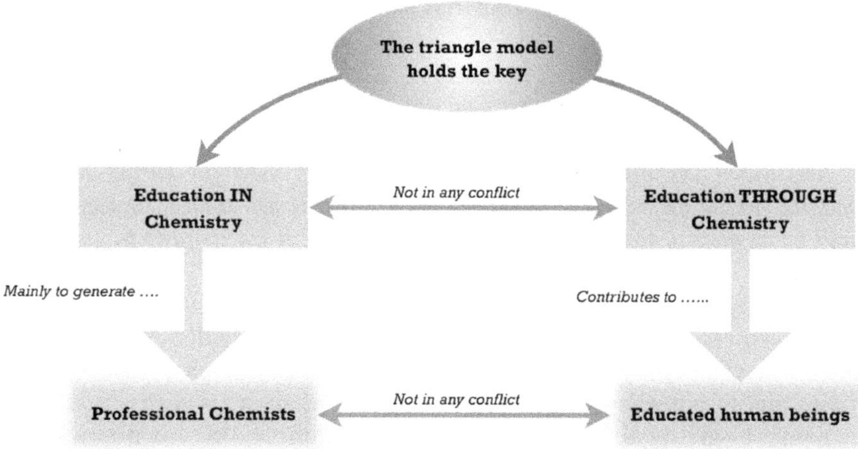


**Figure 4.11** The centrality of triangle thinking.

## 4.9 Chemistry and Assessment

One of the sad features of modern education is the way assessment is now *controlling* what goes on in schools. The pressures on teachers are enormous. School inspections often force them to conform to systems and programmes that, in their heart of hearts, they know are not the best for young learners. As soon as an outsider enters a teaching area, the dynamics of the learning are altered and what the outsider observes is not the reality of what normally happens. Scientists are well aware that the actual '*measurement*' can alter what is being measured. Some teachers have learned to be subversive and '*put on a show*' that they know will please the visiting inspection team (sometimes with the students colluding and cooperating!). They then revert to sounder practices once the inspection is over.

In her insightful analysis of the specious nature of this kind of quality assurance, Danielson exposes the folly of traditional ways of inspection and develops much sounder ways forward, ways that will command the respect of the teaching profession *and* the wider public (Danielson, 2001, 2007, 2008). Her work needs careful consideration in our '*accountability-driven*' education world. She conceptualised quality assurance as built around every teacher as both evaluated and as evaluator. Periodic visits from '*outsiders*' to classroom was simply not a relevant part of the agenda.

Another approach is sadly gaining ground today, this involves comparing school examination results. Too often the '*product*' of the education system is defined in terms of examination results. The task of schools is to educate and equip learners for life after education, not simply to get good examination outcomes. Examination outcomes, by their very nature, can only measure a very narrow range of outcomes, often ignoring what is more important (Popham, 1999; Horn and Wilburn, 2013; Meyer and Schiller, 2013). All this is an example of the outworking of what has become known as

Campbell's Law, a principle that has been shown to apply widely in life (Campbell, 1979, p. 85).

> **Campbell's Law (1916–1996)**
> ***Former President, American Psychological Association***
> *The more any quantitative social indicator is used for social decision-making, the more subject it will be to corruption pressures and the more apt it will be to distort and corrupt the social processes it is intended to monitor.*

What this kind of '*high-stakes*' testing has been shown to do is to force teachers to teach to the test, therefore distorting the wider goals for a rich education. Chemistry cannot escape from these insidious pressures. At its worst, highly conscientious teachers have tried to maximise student performance by giving all kinds of support. For example, there is now a growth industry in private tutoring, and, in some countries, this has evolved to a major '*industry*'. This spoon-feeding has some longer-term effects. There is persistent anecdotal evidence that highly qualified university students are not coping with the more independent learning demanded at that level and are dropping out.

> *Spoon feeding in the long run teaches us nothing but the shape of the spoon.*
>
> E. M. Forster

It is an interesting exercise to relate school grades with final university degrees attained, for those who enter university. The correlations are so low that they do not justify our naive faith in the predictive value of school grades. Universities test a much wider range of skills and the young people develop and mature enormously over their time at university. In fact, school grades are a '*snapshot*' at one moment of time of a very narrow range of educational outcomes.

Overall, it is true that the quality of teaching is vitally important but there are numerous different ways to provide quality teaching (Bailey, 2008) and we cannot reduce the complexity of teaching to ticking a set of boxes. Nonetheless, the power of national assessment is enormous, and the teacher cannot get around this so easily. In the current situation, teachers will only teach what will give their students the maximum credit in examinations. Students will focus on what leads to maximum success.

Research has shown repeatedly that what national examinations and international tests measure is built around recall. Examination Boards may protest otherwise, international testing organisations may make grandiose claims, but the evidence is overwhelming. The way the student gains their answers depends on recall skills, no matter how clever examination setters try to be.

This pattern was the sad feature of education that drove Bloom and his colleagues in their day to develop their famous taxonomy (Bloom *et al.*, 1956). Numerous statistical exercises have confirmed it is still true today (*e.g.* Almadani *et al.*, 2012). However, the best evidence comes by talking to the students as they are undertaking assessment tasks. We shall be amazed at the ingenious ways by which they reach answers. All the evidence points the same way.

Chemistry cannot escape this straight jacket. How is the chemistry teacher to develop the full potential of their students and equip them for life beyond school when the assessment system distorts the outcomes? What makes it worse is when schools are compared using examination data. Highly skilled chemistry teachers, seeking to educate the students through the study of chemistry, may be penalised when they have not expended their energies filling the heads of the students with ever more chemical facts that can generate higher examination results.

When international testing results are employed, it becomes even harder. It encourages the curriculum designers to place even more emphasis on ever more content so that the country can be seen to perform better next time. Pressures are placed on teachers to focus tightly on ways that will bring better test outcomes, at the expense of the wider aspects of a rich education. Today, education has been high jacked as a tool to generate economic wealth following an agenda imposed from outside. There have been numerous studies which have looked at the issues, and universally they reveal major weaknesses that undermine the validity of the entire exercise (Gordon and Reese, 1997; Gardner, 1999; Wolf, 2002; Stobart, 2008; Horn and Wilburn, 2013; Meyer and Schiller, 2013; Trohler, 2013; Stobart, 2014; Ravitch, 2016).

## 4.10 Key Messages

Johnstone's argument never included suggestions that chemistry at the macro level was all that was suitable for most of the school population, leaving the other two levels mainly for those pursuing more academic and more scientifically orientated courses. He wished for an education *through* chemistry as well as an education *in* chemistry, *for everyone*. This is part of the development of the future population, equipped to cope with the scientific issues of the future. This perspective is reflected in all the textbooks he wrote and can be seen in an early paper (Hadden *et al.*, 1974).

He argued that the chemistry curriculum needs to focus on the learners, not the logic of the subject matter. This led to the applications-led curriculum which has been shown to be so successful. Here, macro is centre stage and is rigorous, laying the necessary foundation onto which the sub-micro and representational can be built.

He also argued that we need to be much more careful in the ideas and models we teach in order to make chemistry accessible. He rejected the octet rule, orbiting electrons, atoms '*wanting*' to gain or lose electrons, the dichotomous nature of bonding as covalent or ionic, and the assumption

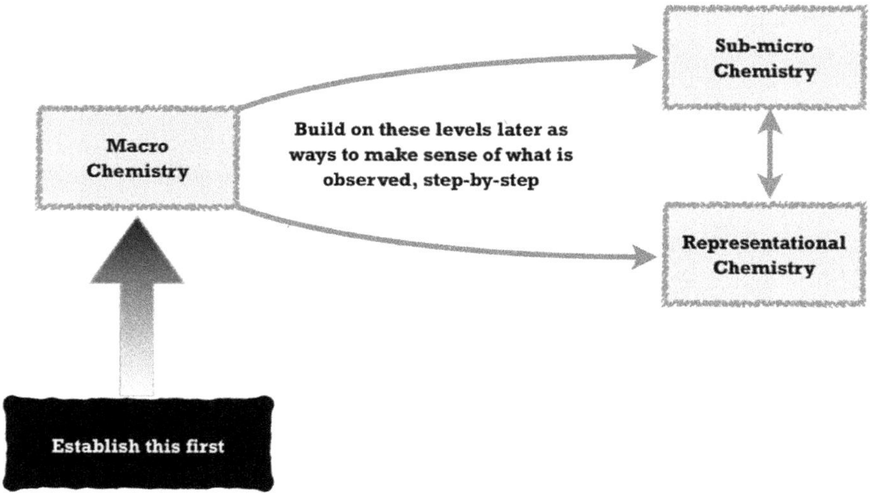


**Figure 4.12** Johnstone's argument.

that we must start with isolated atoms in the gas phase in order to develop principles of bonding and molecular geometry. Much here generates confusion and can led to all kinds of misconceptions and, inevitably, ideas often have to be untaught at later stages.

His argument was simple. Working memory does not allow '*novice*' learners to cope with the three levels *at the same time.* That is just as true for undergraduates as it is for 12-year-old school students. Therefore, we develop the macro level, with rigorous chemistry, before moving into the sub-micro and representational. These can be seen as human attempts to make sense of chemistry, explaining something of *why* the molecular world behaves as it does.

Johnstone (Johnstone and Wood, 1977) placed the laboratory right at the centre of all his teaching (and his books reflect this very clearly). However, he never argued that the laboratory was vital to develop bridges between the three levels. Johnstone was highly creative, but he never suggested that physical models would necessarily create the bridge between the three levels.

His central argument might be summarised (Figure 4.12).

If laboratory work helps in developing the sub-micro and representational, that is great. If models assist in doing the same, that is also great. If discussion groups help, we use them. If re-writing textbooks helps, then we re-write them.

In the next chapter, we shall look at some of these ways forward.

## References

Alenezi D., (2008), *A study of learning mathematics related to some cognitive factors and to attitudes*, PhD thesis, Glasgow: University of Glasgow, http://theses.gla.ac.uk/333/, Accessed 7 November 2019.

Ali A. A. and Reid N., (2020), *Making Sense of Learning, A Research-based Approach*, Berlin: Springer.

Almadani K., Reid N. and Rodrigues S., (2012), What examinations test, *Problems of Education in the 21st Century*, vol. 1, pp. 6–19, http://www.scientiasocialis.lt/qiic/files/pdf/Almadani_Vol.1_qiic.pdf. Accessed 11 December 2019.

Bailey P. D., (2008), Should 'teacher centred teaching' replace 'student centred learning'? *Chem. Educ. Res. Pract.*, **9**(1), 70–74.

Blick D., (1955), The purpose and character of laboratory instruction, *J. Chem. Educ.*, **32**, 264.

Bloom B. S., Engelhart M. D., Furst E. J., Hill W. H. and Krathwohl D. R., (1956), *Taxonomy of Educational Objectives: the Classification of Educational Goals. Handbook I: Cognitive Domain*, New York: David McKay Company. Inc., 7th edn, 1972.

Campbell D. T., (1979), Assessing the impact of planned social change, *Eval. Program Plann.*, **2**(1), 67–90.

Carnduff J. and Reid N., (2003), *Enhancing Undergraduate Chemistry Labor-atoryoratories*, London: The Royal Society of Chemistry.

Chandi S. S., Reid N., McWilliam A. and Gray D., (2009), Exploring the usefulness of a systems-based model from the perspective of biology students, *Scott. Educ. Stud.*, **41**(1), 68–85.

Chu Y.-C. and Reid N., (2012), Genetics at school level: addressing the difficulties, *Res. Sci. Technol. Educ.*, **31**(1), 1–25.

Danielson C., (2001), New Trends in Teacher Evaluation, *Eval. Educ.*, **58**(5), 12–15.

Danielson C., (2007), *Enhancing Professional Practice: A Framework for Teaching*, Alexandria VA: Association for Supervision and Curriculum Development, .

Danielson C., (2008), *Handbook for Enhancing Professional Practice : Using the Framework for Teaching in Your School*, Alexandria, VA, USA: Association for Supervision and Curriculum Development.

Danili E. and Reid N., (2004), Some Strategies to improve performance in school chemistry, based on two cognitive factors, *Res. Sci. Technol. Educ.*, **22**(2), 203–226.

Gardner H., (1999), *Intelligence Reframed, Multiple Intelligences for the 21st Century*, New York: Basic Books.

Gardner J., Harlen W., Hayward L., Stobart G. and Montgomery M., (2010), *Developing Teacher Assessment*, Maidenhead: Open University Press.

Gilbert J. K. and Treagust D., (2009), Towards a Coherent Model for Macro, Submicro and Symbolic Representations in Chemical Education, in Gilbert, J. K. and Treagust, D. (ed.), *Multiple Representations in Chemical Education*, Dordrecht: Springer, pp. 332–350.

Gordon S. and Reese M., (1997), High Stakes Testing; worth the price? *J. Sch. Leadersh.*, 7, 345–368.

Hadden R. A. and Johnstone A. H., (1982), Primary School Pupils' Attitude to Science: The Years of Formation, *Eur. J. Sci. Educ.*, **4**(4), 397–407.

Hadden R. A. and Johnstone A. H., (1983a), Secondary School Pupils' Attitude to Science: The Year of Erosion, *Eur. J. Sci. Educ.*, **5**(3), 309–318.
Hadden R. A. and Johnstone A. H., (1983b), Secondary School Pupils' Attitude to Science: The Year of Decision, *Eur. J. Sci. Educ.*, **5**(4), 429–438.
Hadden R. A., Handy J. and Johnstone A. H., (1974), Education Through Chemistry, *Educ. Chem.*, **11**, 206.
Hanif M., Sneddon P. H., Al-Ahmadi F. M. and Reid N., (2008), The perceptions, views and opinions of university students about physics learning during undergraduate laboratory work, *Eur. J. Phys.*, **30**, 85–96.
Hassan A. K., Hill R. A. and Reid N., (2004), Ideas Underpinning Success in an Introductory Course in Organic Chemistry, *Univ. Chem. Educ.*, **8**, 40–51.
Hofstein A. and Lunetta V., (1982), The role of the laboratory in science teaching, Neglected aspects of research, *Rev. Educ. Res.*, **52**, 201–217.
Holroyd C. and Harlen W., (1997), Primary teachers' understanding of concepts of science: impact on confidence and teaching, *Int. J. Sci. Educ.*, **19**(1), 93–105.
Horn J. and Wilburn D., (2013), *The Mismeasure of Education*, Charlotte, NC: Information Age Publishing.
Hussein F. and Reid N., (2009), Working Memory and Difficulties in School Chemistry, *Res. Sci. Technol. Educ.*, **27**(2), 161–186.
Johnstone A. H., (2000a), Teaching of Chemistry – Logical or Psychological? *Chem. Educ. Res. Pract. Eur.*, **1**(1), 9–15.
Johnstone A. H., (2000b), You Can't Get There from Here, *J. Chem. Educ.*, **87**(1), 22–29.
Johnstone A. H. and Al-Shuaili A., (2001), *Univ. Chem. Educ.*, **5**(2), 42–51.
Johnstone A. H. and Letton K. M., (1988), Is Practical Work Practicable? *J. Coll. Sci. Teach.*, **18**(3), 90–92.
Johnstone A. H. and Letton K. M., (1990), Investigating Undergraduate Laboratory Work, *Educ. Chem.*, **27**(1), 9–11.
Johnstone A. H. and Letton K. M., (1991), Practical Measures for Practical Work, *Educ. Chem.*, **28**(3), 81–83.
Johnstone A. H. and Morrison T. I., (1964), *Chemistry Takes Shape*, London: Heinemann Educational, vol. 1.
Johnstone A. H. and Morrison T. I., (1966), *Chemistry Takes Shape*, London: Heinemann Educational, vol. 3.
Johnstone A. H. and Webb G., (1977), *Energy, Chaos and Chemical Change*, London: Heiemann. Also in Italian: Energia, Caos e Reazioni Chemiche, (translated by L Cardellini and P Mirone), 2002, Piccin, Padova, ISBN 88-299-1630-7.
Johnstone A. H. and Wham A. J. B., (1979), A Model for Undergraduate Practical Work, *Educ. Chem.*, **16**, 16–17.
Johnstone A. H. and Wham A. J. B., (1980), A case for Variety in Practical Work, *Sch. Sci. Rev.*, **61**(217), 762–765.
Johnstone A. H. and Wham A. J. B., (1982), Demands of Practical Work, *Educ. Chem.*, **19**(3), 71–73.

Johnstone A. H. and Wood C. A., (1977), Practical Work in its own Right, *Educ. Chem.*, **14**, 11.

Johnstone A. H., Morrison T. I. and Reid N., (1981), *Chemistry About Us*, London: Heinemann (reprinted 1982, 1986, 1988).

Johnstone A. H., Sleet R. J. and Vianna J. F., (1994), An information processing model of learning: its application to an undergraduate laboratory course in chemistry, *Stud. High. Educ.*, **19**(1), 77–88.

Johnstone A. H., Watt A. and Zaman T. U., (1998), The students' attitude and cognition change to a physics laboratory, *Phys. Educ.*, **33**(1), 22–29.

Jung E.-S. and Reid N., (2009), Working Memory and Attitudes, *Res. Sci. Technol. Educ.*, **27**(2), 205–224.

Mbajiorgu N. M., Reid N. and Ezeano C. A., (2017), *Handbook of Science Education*, Enugu: ESUT Press.

Meyer H.-D. and Schiller K., (2013), Gauging the role of non-educational effects in large scale assessments, in Meyer H.-D. and Benavot, A. (ed.), *PISA, Power and Policy*, Oxford: Symposium Books, pp. 207–224.

Phillips M., (1981), Early history of physics laboratories for students at the college level, *Am. J. Phys.*, **49**, 522–527.

Phillips M., (1983), Laboratories and the rise of the physics profession in the nineteenth century, *Am. J. Phys.*, **51**, 497–503.

Popham W. J., (1999), Why Standardized Tests Don't Measure Educational Quality, *Educ. Leadersh.*, **56**, 8–16.

Ravitch D., (2016), *The Death and Life of the Great American School System: How Testing and Choice are Undermining Education*, New York: Basic Books.

*Chemistry for the Certificate of Sixth Year Studies, Memorandum 4*, Reid N. (ed.), (1982), Dundee: Scottish Curriculum Development Service, pp. 1–106.

Reid N., (1999), Towards an Application-Led Curriculum, *Staff Educ. Dev. Int.*, **3**(1), 71–84.

Reid N., (2000), The Presentation of Chemistry: Logically Driven or Applications Led? *Chem. Educ.: Res. Pract.*, **1**(3), 381–392.

Reid N. and Shah I., (2007), The role of laboratory work in university chemistry, *Chem. Educ. Res. Pract.*, **8**(2), 172–185.

Reid N. and Shah I., (2010), The Idea of the Paper Laboratory, *J. Sci. Educ.*, **11**(10), 8–12.

Reid N. and Skryabina E., (2002a), Attitudes Towards Physics, *Res. Sci. Technol. Educ.*, **20**(1), 67–81.

Reid N. and Skryabina E., (2002b), Gender and Physics, *Int. J. Sci. Educ.*, **25**(4), 509–536.

Sellinger B., (1983), *Chemistry in the Marketplace*, Australia: Harcourt Brace Javonivich Group.

Shah I., Riffat O. and Reid N., (2007), Students Perceptions of Laboratory Work in Chemistry at School and University in Pakistan, *J. Sci. Educ.*, **8**(2), 75–78.

Shell D. F., Brooks D. W., Trainin G., Wilson K. M., Kauffman D. F. and Herr L. M., (2010), *The Unified Learning Model*, Dordrecht: Springer.

Sirhan G. and Reid N., (2001), Preparing the Mind of the Learner - Part 2, *Univ. Chem. Educ.*, **5**(1), 52–58.
Sirhan G., Gray C., Johnstone A. H. and Reid N., (1999), Preparing the Mind of the Learner, *Univ. Chem. Educ.*, **3**(2), 43–46.
Snow C. P., (1959), *The Two Cultures and the Scientific Revolution*, Cambridge: Cambridge University Press.
Stobart G., (2008), *Testing Times: The Uses and Abuses of Assessment*, London, New York: Routledge.
Stobart G., (2014), *The Expert Learner, Challenging Myths of Ability*, Berkshire: Open University Press, McGraw-Hill Education.
Trohler D., (2013), The OECD and cold war culture: thinking historically about PISA, in Meyer, H.-D. and Benavot A. (ed.), *PISA, Power and Policy*, Oxford: Symposium Books, pp. 141–162.
Tsaparlis G., (2009), Learning at the Macro Level: The Role of Practical Work, in Gilbert J. K. and Treagast D. (ed.) *Multiple Representations in Chemical Education*, Springer, pp. 109–136.
Ud-Din M. N., Reid N. and Malik I. A., (2016), *Training Manual on Educational Assessment*, Islamabad: Federal Board of Intermediate and Secondary Education.
Venville G., Wallace J., Rennie L. and Malone J., (2002), Curriculum integration, *Stud. Sci. Educ.*, **37**, 43–83.
Wadsworth B. J., (1979), *Piaget's Theory of Cognitive Development: An Introduction for Students of Psychology and Education*, London: Longman.
Wolf A., (2002), *Does Education Matter? Myths About Education and Economic Growth*, London: Penguin.

CHAPTER 5

# *Johnstone's Triangle and the Learning Process: Making Understanding Accessible*

## 5.1 Looking Back

In the last chapter, we saw that Johnstone's central argument was that we start with the macro chemistry, which can be quite rigorous in nature. When that is well established, then we carefully build on the sub-micro and representational. These can be seen as ways by which we are attempting to interpret and explain what is observed at the macro level about the way the molecular world behaves.

He saw chemistry as an important element in education, relevant and important for all learners. The sub-micro and representational are not just for the academic elite. Indeed, he pioneered many ways by which both the sub-micro and representational might be made accessible for all learners. However, Johnstone (2000) argued strongly that we need to take the psychology of the learner into account and not determine our curricula and teaching on the basis of the logic of chemistry as an academic discipline. The logic of chemistry as an academic discipline has developed from centuries of research and is our '*expert*' attempt to impose a framework of understanding on a vast range of experimental observations. The framework comes after very considerable experience of the observations.

This is where the psychological dimension is so important. The '*novice*' learner lacks this experience of the macro. The framework can make complete sense to the teacher but may make little sense to the learner. Johnstone argued that the macro must be placed centre stage and it can be rigorous.

Advances in Chemistry Education Series No. 6
The Johnstone Triangle: The Key to Understanding Chemistry
By Norman Reid

Published by the Royal Society of Chemistry, www.rsc.org

It lays the necessary foundation on to which the sub-micro and representational can be built.

There are numerous ways in which the sub-micro and representational can be introduced step-by-step and we shall look at some of the approaches that have been found to be useful.

## 5.2 The Role of Language

It is self-evidently obvious that ideas and understandings are communicated mainly by means of language. However, there is also extensive research on the way understandings are gained by means of the visual–spatial. It has to be stressed that the visual–spatial is not referring to any artistic ability to draw. The visual–spatial is better to be seen more in terms of *relationships* between ideas rather than following the logic of a presentation or argument.

In some ways, thinking in this visual–spatial way is natural. Humans are seeking to make sense of the world around them. That inevitably involves ideas being brought together. Ideas are related to each other, in an attempt to make some kind of coherent whole. The world of the logical and the sequential may well be less natural. We shall look further at the visual–spatial in a moment but let us return to language.

Chemistry, like any other discipline, has its language and terminology. It is often thought that this specialised language causes problems for learners. Early in his career, Johnstone explored the central role of language and found that it was not the technical language of chemistry that was the main source of problems. It was the way chemistry employs ordinary English words, but invests them with specific meanings. These may or may not reflect the meanings used in wider communications.

In a series of publications, Cassels and Johnstone (1977a, 1977b, 1978a, 1978b, 1980, 1983, 1984) presented the evidence that showed how language could hinder (or even confuse) understanding and how language could undermine assessment validity. Learners could cope with the specialised language not too badly. They coped much less well with ordinary English words which were being used to carry very specific meanings in chemistry. Examples include words like: '*abundant*', '*crude*', '*energy*', '*feasible*', '*isolate*', '*liberate*', '*pungent*', '*residue*', '*theory*', '*valid*' and '*volatile*'. The problem words were listed in a booklet (Cassels and Johnstone, 1985).

The clear message is that we need to make it very clear to learners how such words are being used. The message needs repetition so that working memory is not taken up each time by the learner in interpreting such words in the specific way they are used in chemistry. As teachers, we need to be acutely aware of multiple meanings that some words and phrases carry. We need to spell out how specific words are being employed and the meaning they carry in chemistry, distinguishing these carefully from other wider meanings.

However, there is another side to language. English is now the language of the sciences. Much chemistry is taught in many countries using English. This is not the first language of the learners. We noted in Chapter 2 that working in a second language, even when competent in that language, employs part of the working memory capacity for '*interpretation*'. It is highly likely that something similar is happening when chemistry employs ordinary English words in a specialised way: part of the part of the working memory capacity is employed for '*interpretation*'.

There are some straightforward messages for us here. '*Novice*' learners need the maximum working memory capacity available to them simply to gain understanding. Not only must we avoid working at all three levels at the same time, but we must also avoid language that is unnecessarily complex or is confusing because chemistry considers a specialised meaning that does not match the '*normal*' usage of the word. Language can take up precious space in the working memory, leaving less for understanding.

**Abstraction–Erudition**

One sad feature of much educational literature is the way simple ideas are presented in such complex language.

***Abstraction is confused with erudition.***

In presenting chemistry, the ideas can be complex, and we need to learn how to present them in simple clear language.

***Working memory dictates this.***

However, despite the central importance of language, there is now considerable research showing how learners often move into the visual–spatial. Here is an area where much more research is required. We need to explore how the general findings relate specifically to the world of understanding chemistry. Johnstone's triangle may offer a key starting point for our thinking in this area.

## 5.3 The Visual–Spatial

In Chapter 2, we noted the increasing interest in the realm of the visual–spatial, especially in relation to the operation of working memory. Silverman (1989, 2002) has brought this to the attention of the wider educational world. Hindal has explored this in relation to performance across all subject areas (Hindal, 2007; Hindal *et al.*, 2009; Hindal *et al.*, 2013; Hindal, 2014).

Chemistry is a highly logical subject. The evidence suggest that a very high proportion of the population naturally think in terms of relationships rather than in a logical series of arguments or stages of thought. We are so fond of our bullet points and lists. Many learners think more in terms of how ideas connect to each other.

**Understanding**

This involves the linking of ideas together to form a coherent whole. Linear logic can often make that demanding.

David Ausubel (1968) knew the importance of this when he laid so much stress on knowing what learners already know and understand, before presenting new ideas.

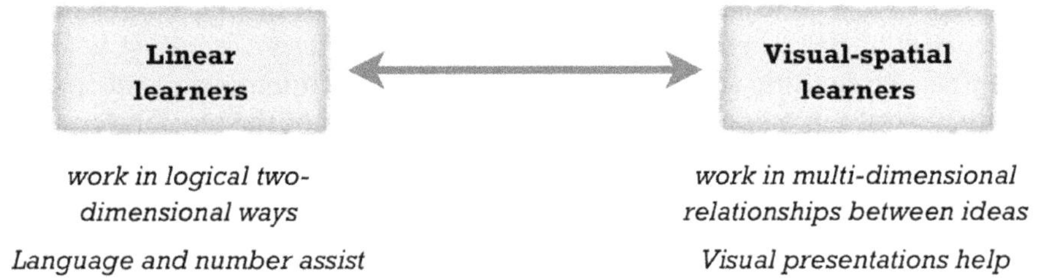


**Figure 5.1** Different learning approaches.

The nature of visual–spatial thinking is illustrated in Figure 5.1.

Research has shown that those who are strong in the visual–spatial tend *always* to perform better in all subject areas (Hindal, 2007; Hindal *et al.*, 2009; Hindal *et al.*, 2013; Hindal, 2014). This makes sense in that, if we can group ideas together to be seen as one relationship, then this is a form of chunking (Miller, 1956). Is the way we present chemistry hindering their understanding of that chemistry? We saw in Chapter 1 that the working memory has a visual–spatial function: what Baddeley called a visuo-spatial scratchpad (Baddeley *et al.*, 2015). However, as Hindal pointed out, if the working memory is able to store understandings in a more visual form, then they can be handled as one item, reducing the likelihood of working memory overloading so easily.

> **Semantic space**
> There is a considerable literature that looks at how we see meaning and employs ideas related to '*semantic space*'.
> The concept was pioneered with the work Charles Osgood (1916–1991) (Osgood *et al.*, 1957).
> [Discussed in chapter 3 of Ali and Reid, 2020.]

In the context of chemistry, a mathematical relationship can bring together several ideas and act as a chunking device. Therefore $PV=nRT$ or $\Delta G^{\circ}=nFE^{\circ}$ can both act as chunking devices. Several variables can be brought together and seen as one. Similarly, chemical formula and equations can act as chunking devices. They are bringing together several ideas and they can be seen as one, reducing potential working memory overload. However, for this to happen, the relationships or equations *must hold meaning*. If the learners are struggling with the meaning of the variables and are not confident in understanding formulae, then the power to chunk is lost (Figure 5.2).

Chemistry is full of concepts. Concepts are only understood when the underpinning ideas can be held in working memory *at the same time*. However, once grasped, mathematical relationships, chemical formula or chemical equations, diagrams or models can act to summarise the relationships and to help to enable the concept to be seen as one. There is then

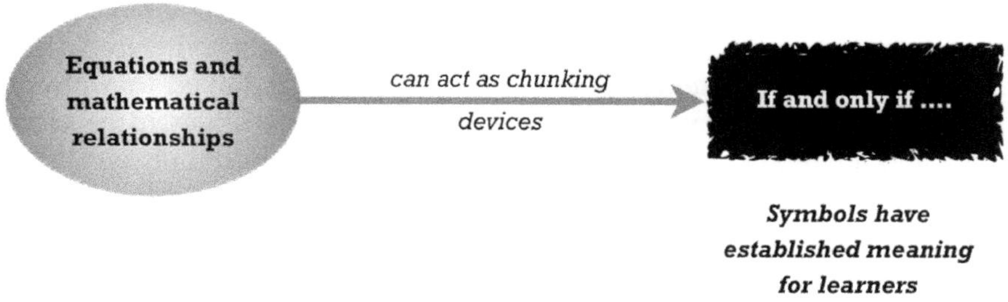


**Figure 5.2** Chunking devices.

working memory capacity left to think about the concept, its meaning and how it can be employed.

There is a fascinating example of this in medical education (Johnstone and Otis, 2006) where clear evidence was gained that suggested that the more able students were storing ideas as chunks and then seeing them as chunks in later work. In passing, the work employed mind-mapping approaches and it was found that the *more able* students developed *simpler* mind-maps, simply because they were chunking more efficiently. This illustrates the dangers in trying to use mind-mapping as an assessment tool.

## 5.4 Diagrams and Physical Models

Bucat and Mocerino (2009) discuss the use of structural representations, noting the problem of employing two-dimensional diagrams on paper to represent three-dimensional reality. They stress the importance of the models employed being accessible to students and not generating even more misconceptions now or in the future. They also note that so often our representations deal with individual entities, leaving the '*novice*' with difficulties in perceiving that the real world involves huge numbers of entities. However, there are other issues: the representation must be faithful to the reality as we currently know it and must be accessible to the learner: able to be comprehended *within* the linked capacity of working memory.

**Misleading Models**

Enormous creativity has been employed in taking science out to young children by means of outreach activities, visiting speakers and, particularly, science centres.

It is salutary experience to survey these activities and ask serious questions about the models employed and the extent to which they may well develop misconceptions.

The use of colour can help. The use of computer representations can help. One of the best answers is the use of physical models that learners can actually handle. Many years ago, Johnstone developed a simple way to link the two-dimensional to the three-dimensional. He set-up an area in the chemistry department in Glasgow where there was a series of chemical tasks, set out as '*stations.*' Students could work through these when convenient to themselves. At each '*station*', a printed worksheet was linked to a physical model of a chemical entity and there were also a few slides to be viewed down a hand viewer. The student could then move from the physical three-dimensional model across to the slides (which showed the same structure in colour) and then link both to the drawing on paper. In undertaking the set tasks, the students were continually moving between physical models, paper representations, with the coloured sides sometimes acting as a kind of link.

Davidowitz and Chittleborough (2009) note that the representational can form a link between the macro and the sub-micro. The goal is the development of mental pictures and models which accurately reflect what is known as it is interpreted at the sub-micro level. Inevitable, diagrams tend to be static while chemical systems are dynamic. However, there is a fundamental issue with all physical models and all diagrams. The models and diagrams themselves may place considerable pressure on limited working memory resources. We may end up not helping students at all but generating even greater problems.

Table 5.1 summarises some of the more significant problems posed by the use of models and diagrams.

Johnstone never argued for the use of models or diagrams in relation to the three levels and he never suggested that either models or diagrams could be used to link the levels. However, he did use diagrams and he invented numerous physical models. Models or diagrams may be of value in linking levels or they may generate further problems. We need to be careful not to assume that, inevitably, they hold the answers (Figure 5.3).

**Table 5.1** Problems with Models.

| Potential Problem | Comment |
|---|---|
| *Dimensions* | Chemical entities involve three-dimensions. Diagrams on paper have limitations. Colour or computer graphics can help. Physical models can help. |
| *Single entities* | Chemistry involves entities at the mole level, but we can only draw single entities or very small numbers. |
| *Static* | Chemical situations are almost always dynamic. While computer animations can help, they can also mislead or overload the working memory. |
| *Reality of their own* | Models and diagrams have a kind of permanence and can create a reality of their own which do not reflect the actual reality perfectly. |
| *Working memory overload* | This is a big problem, largely ignored in the literature on the use of models in chemistry. Chemical situations are complex, and models and diagrams may require more working memory capacity than is available. |
| *Generate misconceptions* | Inadvertently, models can generate mental models that lead to misconceptions, causing later confusions or the need for '*unteaching*'. |

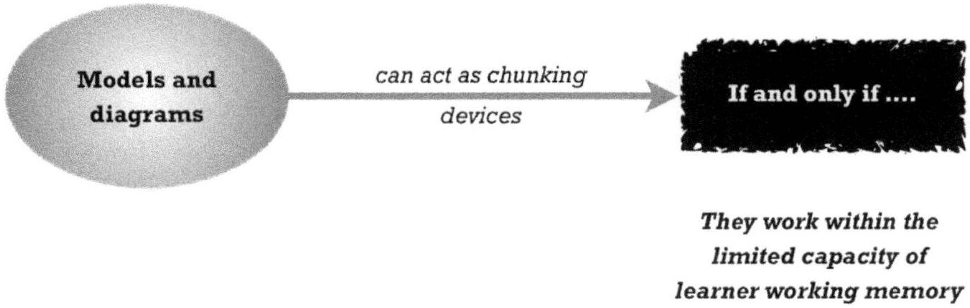


**Figure 5.3** Models and diagrams.

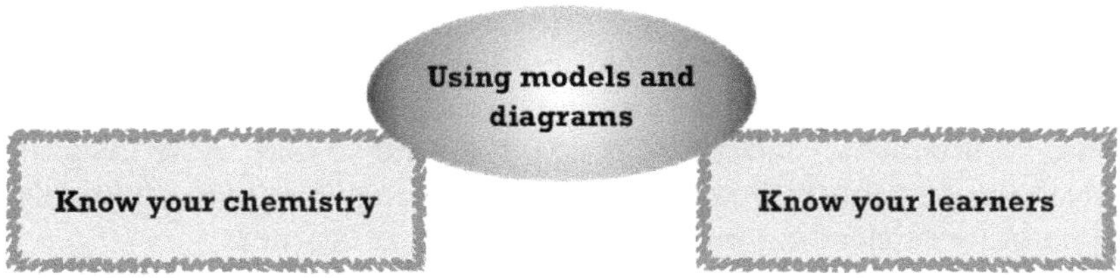


**Figure 5.4** Using models and diagrams.

The potential in developing and using models (physical or drawn) or diagrams is that they open the door for learners to be able to build mental models which can represent what is thought to happen at the molecular level. Whether models can link the three levels outlined by Johnstone is simply unknown, but it does make some sense. In developing any model or diagram in chemistry, there are two important aspects (Figure 5.4).

It is essential that we really understand the relevant chemistry in order to avoid creating confusions or generating potential misconceptions. Therefore, using target models to show electron sharing or electron transfer may look elegant, but it is so far away from reality that it will utterly mislead. It generates later problems when it is seen not to be correct.

**Planetary models of atoms**

It is tempting to argue that these are simple and, therefore can help learners at early stages. They then can be modified later.

In fact, the evidence shows that the Johnstone approach, which avoided any use of planetary models, is far more accessible for it links the sub-micro understanding to macro phenomena. The great advantage is that later un-teaching is avoided.

His approach leads naturally on to more complete understandings while the planetary model cannot be adapted: it is simply wrong!

Johnstone sometimes casually referred to teaching in chemistry as a process of diminishing deceptions. However, the same issue has been taken up cogently by Ball when he talks of '*Lies to Children*' (Ball, 2020). He notes

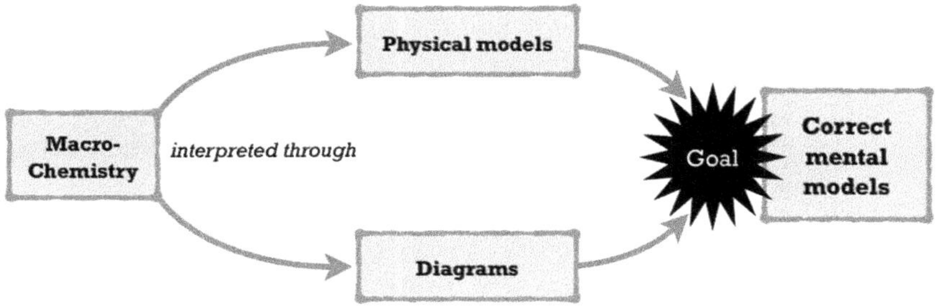


**Figure 5.5** The difficult goal of creating correct mental models.

the often-repeated argument that the truth can be given later. However, the evidence shows that '*unlearning is often very difficult*'. Specifically, he challenges the use of the solar system atom, the octet rule or the '...*the unrealistically neat divisions into ionic and covalent bonding or how oxidation states are introduced*' (Ball, 2020, p. 17). We need to take these insights seriously in our current textbooks and in our teaching. We are in serious danger of generating misleading models in the minds of learners.

It is also essential that we know our learners. We need to keep right at the forefront of our thinking that their working memory capacities are fixed and limited. Our students are, in varying degrees, '*novice*' learners. They come to everything we teach with background ideas that may or may not be helpful. We need to understand what they understand and have some idea how they came to that understanding. There is no short cut to this. Experienced practising teachers hold the key for they are best placed to understand the students they teach.

For most of his career, Johnstone was active in teaching chemistry. Much of his research grew out of this experience. He never lost touch with the learners and was an observant and good listener. His development of physical models grew out of his experiences in seeing where learners *actually* faced problems in understanding. He was acutely aware of the role of working memory. The models he developed offer some useful practical examples which still have an important place for us today. We shall look at a few of them. They hold the potential to aid learners as they develop their own mental models of what is happening at the molecular level.

There is a logic to all this which Johnstone understood very clearly (Figure 5.5).

## 5.5 Models and Molecular Structures

One of great ways that Johnstone introduced was the use of modelling balloons to represent molecular structures. The foundations were laid in a textbook for 12-year-olds when he established the idea that like charges repel and opposite charges attract. To illustrate this, he used ordinary party balloons (Johnstone and Morrison, 1964, p. 55). This was used to offer some kind of rationale behind the conduction of saltwater and the electrolysis of water. Little was made of it that stage, but the idea was established. The idea was touched on in looking at radioactivity which, historically, offered clues about the structure of atoms (Johnstone and Morrison, 1965, pp. 42–57). The next two books developed the ideas further, with balloon models illustrating aspects of halogen chemistry before really coming into their own in carbon chemistry (Johnstone and Morrison, 1967, pp. 24–25). In the final book of the series, balloons again re-appear in more formal carbon chemistry for 17-year-old learners (Johnstone and Morrison, 1969, pp. 43–45).

The way ideas keep re-appearing is sound pedagogy, reflecting the '*spiral curriculum*' argument developed by Jerome Bruner (Bruner, 1960). Here, topics and ideas are re-visited many times, each successive visit adding a new insight and enriching the original idea. This is consistent with the triangle model and the underlying basis that limited working memory capacity *controls* extent of understanding. Johnstone started with macro and then, step-by-step, themes were revisited, and sub-micro and representational insights added. In the final volume of this series of texts, it is interesting to see that the three dimensional understanding of organic molecules, represented by modelling balloons is portrayed alongside the two-dimensional way such molecules are typically represented as they are '*flattened*' onto the printed page (Johnstone and Morrison, 1969, pp. 43–45).

In his later book, the balloon model approaches developed further (Johnstone *et al.*, 1981). The students already knew the basic ideas of attraction and repulsion of electric charges (having met that previously in both physics and chemistry). They had met the words '*proton*' and '*electron*'. He then asked the simple question: if we have two or more atoms close together, how the negative electrons might behave. He looked at molecular situations and did not start with isolated atoms in the gas phase!

Using balloon models, he pictured two adjacent positive nuclei, being held together by a concentration of electrons between them (Figure 5.6). The concept of the chemical bond was therefore introduced in meaningful terms.

This model is built on established ideas relating to charge attraction. It does not assume electron pairing, nor any ideas of '*stable octets*'. Able students often find the idea of electrons being required to pair up as instinctively unlikely (they are still both negatively charged) while there is no obvious reason for learners to understand any reasons why attaining octets is desirable.

What happens when one nucleus is caught between two others (Figure 5.7)?

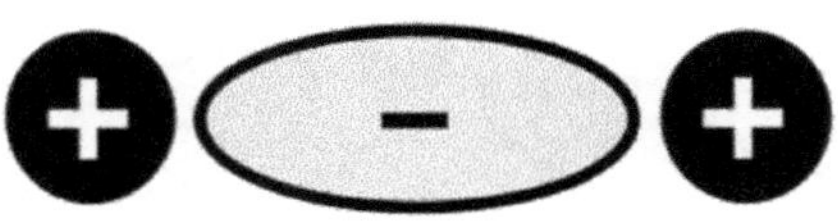

**Figure 5.6** The concept of the chemical bond.

**Figure 5.7** The linear molecule.

At this point, the usefulness of modelling balloons really comes into its own. He asked about the shape of the molecule formed when one atom links to three others.

The repulsions between the concentrations of negative charge around the central atom generates a trigonal structure with angles of 120° (Figure 5.8). He then moved on to consider four atoms around one central atom. Looping two modelling gallons through each other at the centre point demonstrates this easily (Figure 5.9). It makes complete sense to the learners.

The students were allowed to handle the balloons. They could *'feel'* the repulsions between the four *'legs'* in the tetrahedral model simply because of

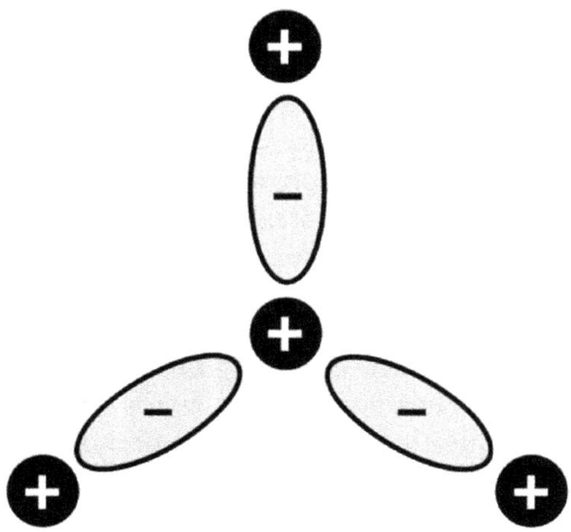


**Figure 5.8** The trigonal molecule.

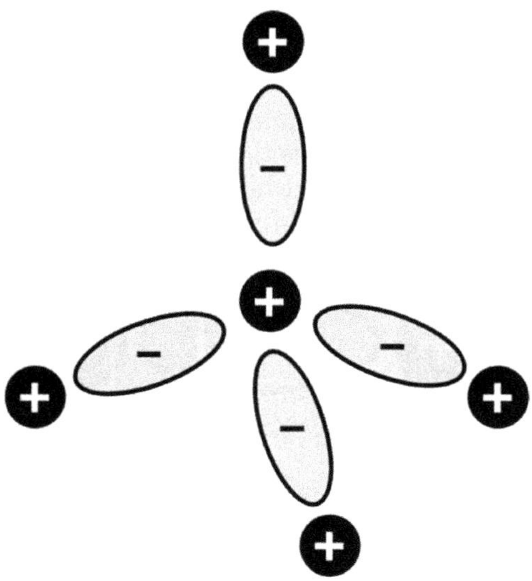


**Figure 5.9** The tetrahedral structure.

the nature of elastic balloons. The entire structural ideas were built on nothing but a simple understanding of attraction and repulsion between electric charges.

At a later stage, with senior school students, the balloon idea was revisited for a final time, to generate octahedral structures. Overall, the lessons were fun. The ideas did not depend on mysterious ideas that lay out with normal experience. The structures arose naturally. There was no recourse to such ideas like hybridisation, which appear weird and unreal.

Later, it was then easy to move ideas forward to structures involving lone pairs (*e.g.* water and ammonia). The stereochemistry of common compounds like sulfur(IV) oxide posed no problems. This structure does not conform to any supposed '*octet rule*' but, if the rule has never been taught, then learners face fewer difficulties in understanding what is happening. '*Unteaching*' is, therefore, avoided.

The idea that non-bonding electrons are slightly closer to the nucleus is easy to understand, they are only attracted by one nucleus. This leads on to the idea that molecular shapes may be distorted from perfect geometry simply by looking at the proximity of electrons in various bonds and non-bonding electrons. The huge advantage of the Johnstone approach is that it is consistent and coherent, never requiring young learners to accept irrational propositions (like atoms '*wanting*' to lose electrons). It starts with simple ideas that are accessible from the start and then builds things up in stages, never having to undo any ideas at later stages.

Multiple bonds could be represented easily, and Figure 5.10 illustrates how a double bond can be conceptualised and why they pull the adjacent nuclei closer together when compared to a single bond.

If modelling balloons are employed, then the stereochemistry of the carbon to carbon double band is illustrated vividly in showing the two other bonds (for example in a molecule like ethene).

The approach never invoked specious ideas like '*octet rules*' or atoms '*wanting*' to achieve octets. Interestingly, it did not assume the bonds were always electron pairs. It just pictured negative change caught between two adjacent positive nuclei, being attracted to both nuclei, therefore holding the nuclei together. With small atoms, it was physically difficult to have more than four other atoms adjacent. With larger atoms, there is more space allowing more complex stereochemistry.

The approach allowed the development of two electron bonds to develop naturally as needed but three electron bonds and delocalisation generated

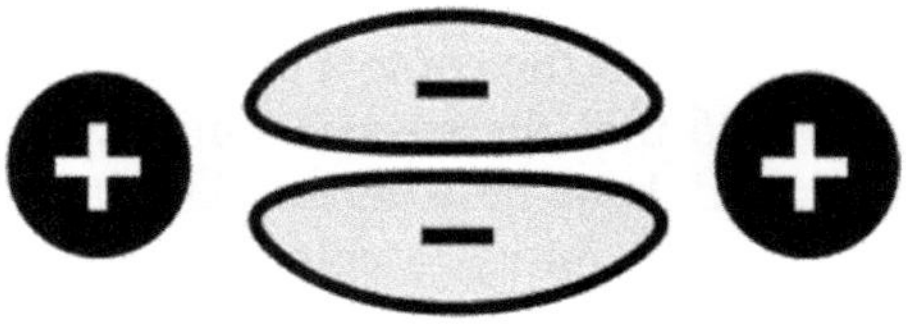


**Figure 5.10** The double bond.

few problems when these ideas were needed. Hybridisation is a mathematical manipulation because we *assume* that the starting point is single atoms in the gas phase. That can happily be left to university chemistry education and presented in this way. For the school student, the idea of isolated atoms and ions in the gas phase is somewhat beyond life experience!

In the final year in Scottish schools at that time, much was built around transition metal chemistry and there was need to conceptualise d-orbitals. So often, d-orbitals are shown as five drawings or five models (there are many excellent drawings online). Working with his technician, Johnstone developed a single model where there were spaces for each of the five orbitals and a place where *all five could be placed around one central point*. The electron clouds were made of wood, each of the five being painted a different colour. When the five orbitals were then superimposed in this way, the spherical symmetry immediately is obvious. You can find a good picture of his model although it is in black and white (Johnstone, 1971). The physical model was marketed for a while by a chemical equipment supplier but now seems to have disappeared.

## 5.6 Moving Models

Molecular situations are not static but dynamic and these are not easy to represent. Johnstone developed a very simple idea to illustrate the dynamic nature of equilibrium. They can be found in one of his books (Johnstone and Morrison, 1969, p. 140). Two students were placed in front of two large glass troughs. One student was given a 100 mL beaker, the other a 400 mL beaker. At the start, one trough contained water, the other was empty.

One student used the beaker to transfer water from the full trough to the empty one and the other to transfer the water back again. To avoid a lot of broken glass, it was not a race: the transfers were kept in step with each other. The alternative is to employ plastic beakers. Eventually, there seemed to be no further changes in depth of the water in the two troughs. At that point, the student with the 400 mL beaker scooped as much water as was possible into the beaker and then poured it into the 100 mL beaker. It was about 100 mL! At equilibrium, the forward and backward reaction are in balance.

The model is by no means perfect, but it does illustrate the dynamic nature of equilibrium and leads naturally on at later stages to the learners understanding ways by which equilibria can be moved in desired directions.

The $S_N2$ mechanism (which was included at that time in the final year school chemistry curriculum in the context of reactions of haloalkanes) was not easy to demonstrate in that the stereochemistry is inverted during the reaction. With the skilled help of a technician, Johnstone developed a mechanical model where one entity (a coloured wooden ball) approached from one side and the stereochemistry moved through planar to the inverted stereochemistry while the opposite atom dropped off.

To illustrate the functioning of a mass spectrometer, Johnstone had steel balls of variable sizes (and, therefore, masses) moving down an inclined slope under which had been attached a magnet. The extent of deflection was seen to relate to their masses. This allowed the principles of mass spectrometry to be introduced at age 14, allowing learners or see how masses of atoms could be compared and how isotopes are identified.

The brilliance of the models Johnstone created was not only their simplicity but also that they aimed to communicate *one idea at a time*: the working memory could cope, and understanding was enhanced. In addition, his models were based on the macro and therefore ideas were accessible to the learners while the way the physical models were constructed reflected the reality of the chemistry situation as closely as possible.

## 5.7 The Behaviour of Molecular Entities

A class of students (often very young: aged about 12) could be invited to be atoms (or molecules). They were acting the part of molecular entities. First, the class was asked to stand along one side of the room in a series of lines parallel to the wall, tightly packed together. One student was then selected and invited to push their way through the lines. The lines held firm!

The class was then invited to remain in their lines but allowed to '*wobble*' about slightly. The single student could now push their way through, with an effort. Finally, the class was then invited to move through the entire laboratory space in the room. Occasionally, they bumped into each other or an item of furniture.

This simple model, using no more than the students, gives an excellent illustration of the behaviour of entities in solids, liquids and gases. One of the common misconceptions is the spacing in liquids will be '*half-way*' between that in gases and solids. This human model helps to clarify this.

Figure 5.11 was used much later by Al-Osaimi (2013, p. 376) in the teaching booklet he developed for 12–13-year-old learners. He was careful to stress that:

> *. . ..no one can see the particles. However, many experiments have suggested the following pictures. In all the pictures, the molecules are shown as tiny circles. Most molecules have much more complicated shapes, but the circle will do for the moment.*

The teacher could then discuss the boiling of water and how much steam emerged for the loss of a small amount of water. This can be followed by looking at the freezing of water to form ice where the volume of the ice formed was not that different from the volume of the water (the anomalous behaviour of water freezing was not considered at this age).

Johnstone had a box made with the shape of a cube, of side approximately 28.2 cm. This gives the gas molar volume of 22.4 L at stp. 18 mL of water

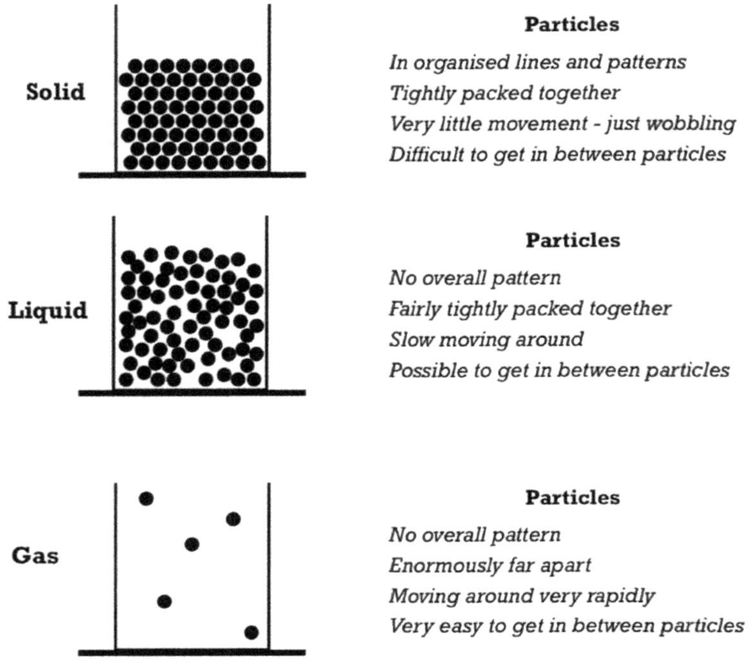


**Figure 5.11** Solids, liquids and gases.

would generate a volume of steam of approximately this volume (ignoring expansion effects). Therefore, this gives the learner a '*feel*' for the spacing in gases and liquids, again all at the macro level.

One of the factors that influences chemical reactions enormously is the actual size of ionic or molecular entities. Johnstone set up four delivery tubes vertically. Tube one contained 0.1 mole of sodium fluoride, tube two 0.1 mole of sodium chloride and so on. The students at this stage knew that the sodium halides were ionic. They knew that the mole was a way of counting entities at the molecular level. Therefore, each tube contained the same number of sodium ions and the same number of halide ions. Here was a simple way to illustrate the growth in ionic sizes going down a column of the periodic chart (Johnstone and Morrison, 1966, p. 2). If we do the same with tubes of equal molar masses of glucose and sucrose, we find that the latter fills a tube almost twice the height of the first, demonstrating how the size of the dimer compares to the monomer.

He developed the same idea looking at a homologous series of alkanols (alcohols) (Johnstone, 2010, p. 26) (Figure 5.12).

This shows an average volume increase of about 16.6 mL. It is evidence, demonstrated at the macro level, of the increasing size of molecules. In this case, it is the addition of one $-CH_2$ group from one alkanol to the next. Here are simple ways to make molecular situations accessible at the macro level.

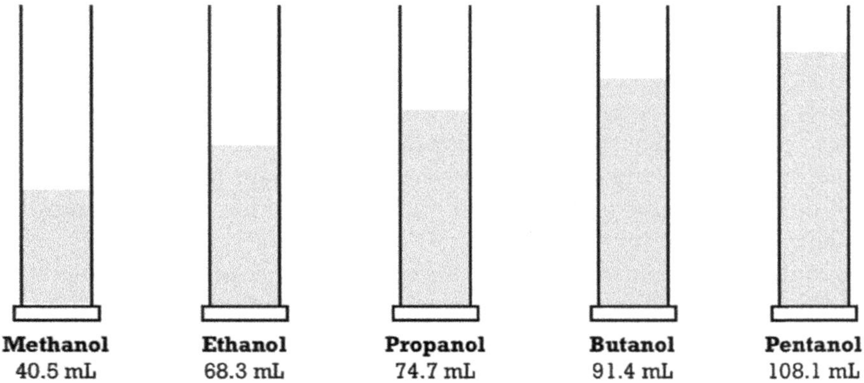


**Figure 5.12** Comparison of molar volumes of a homologous series of alkanols (alcohols).

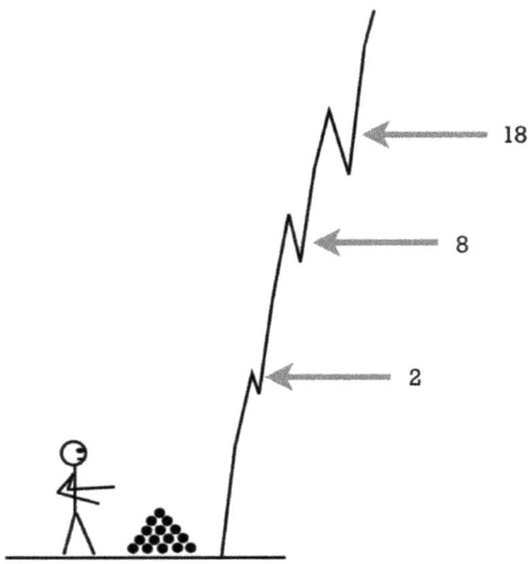


**Figure 5.13** The crazy tennis balls.

At an early stage, the learners were required to consider electrons entering shells, the Aufbau principle. Johnstone started by drawing a very steep slope with small crevices on its face.

The students were asked to imagine the ridiculous situation where they stood at the foot of the steep slope with a very large supply of tennis balls and they were required to throw them up the slope face (Figure 5.13). The balls would catch in the crevices, but the capacity of the crevices took a maximum of 2, 8 or 18 going up. If an extra one landed just above a full crevice, then it just fell back again. The idea of quantisation of energy was implied in the model although this was never followed up this age. Again, a macro situation was employed to illustrate a sub-micro situation. In addition, the sheer ridiculous idea of anyone with a large supply of tennis balls, throwing them up a cliff face gives a story that students would never forget!

It is an interesting challenge to ask the students if they can see any pattern across the numbers: 2, 8, 18, 32. Indeed, it is an interesting challenge to ask them to predict what the 5th number should be (if it existed). Usually, some students work out the gaps between the numbers and then predict the next will be 50. The more mathematical might observe that they are double the square of the first four integers: $2\times(1)^2$, $2\times(2)^2$, $2\times(3)^2$ and $2\times(4)^2$.

From this, the next will be twice 25. This, usually, leads into an interesting discussion. Why? In fact, the numbers arise from solutions to the Schrödinger equation, but it raises issues about the mathematical basis of the physical universe. These strange numbers have meaning. They are not just arbitrary.

## 5.8 Graphs and Group-work

Many curricula have long sections related to trends across the Periodic Table. The students find this excessively boring. The shape of the Periodic Table itself is not always easy to grasp. Johnstone employed the models to illustrate increasing sizes of ions and molecules. However, he also made extensive use of graphs to show patterns, an approach that many have employed since.

For example, in looking at atomic sizes, the graph can be shown as Figure 5.14.

Numerous questions can be discussed arising from the shape of the graph in Figure 5.14. For example,

(a) Identify the elements at the peaks. Can you explain? Why do the peaks become higher from left to right?
(b) Identify the elements at the troughs. Can you explain? Why do the troughs become higher from left to right?

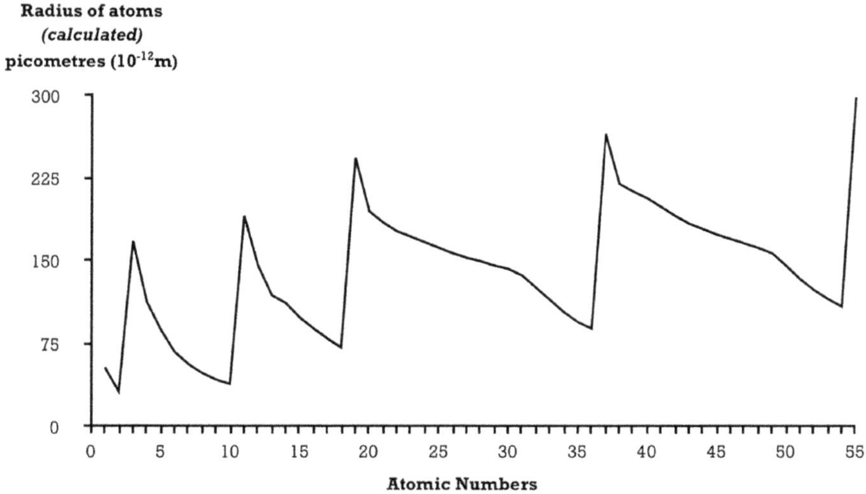


**Figure 5.14** Atomic sizes for the elements.

(c) There are steady falls between each peak then a large jump to the next peak. Can you explain the falls?
(d) Look at the elements at the peaks: how do they behave?
(e) Look at the elements at the troughs: how do they behave?
(f) The falls are not quite linear: can you offer any interpretation?

We can take any trends across the periodic table and present it in graphical form. For example, first ionisation energies can be presented (Figure 5.15). Working in small groups, students can identify the elements at the peaks and troughs and consider how both the peaks and troughs get steadily lower. In the light of what they already know about the elements, they can start to discuss reasons why such patterns exist.

This idea of Johnstone was later taken up by both Hussein and Al-Osaimi. They were developing teaching materials to test understanding and critical thinking skills, respectively. In Hussein's re-presentation of material related to periodicity, students could work on their own or in groups (Hussein, 2006, pp. F5–F56; Hussein and Reid, 2009) while Al-Osaimi (2013, pp. 407–420) was working with younger learners and they were encouraged to work in groups. The key thing is to focus on one aspect of periodicity at a time. Working memory can then cope.

All of this can be taken further at a more descriptive level. In many parts of the world, chemistry is presented in a very abstract way, with limited laboratory work. The entire presentation of periodic trends and patterns is left unrelated to any conception of what the elements actually look like and how they behave. Indeed, it is very easy for the approach to be detached from real life. The sub-micro is then being presented with almost no macro experience on to which it can be built.

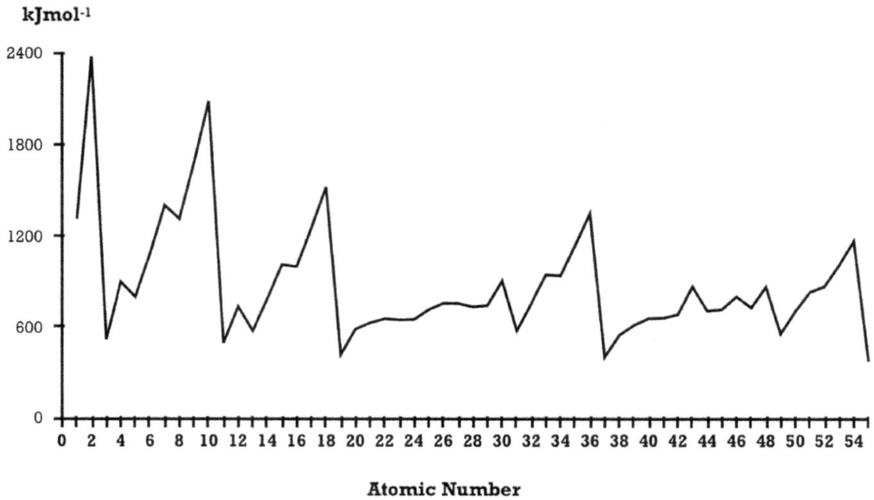


**Figure 5.15** First ionisation energies.

**Table 5.2** Periodic trends.

| Trend | Age | Structure |
|---|---|---|
| Element name plus date of discovery. Element name plus occurrence (coloured coded). | About 11–12 | Students work in small groups to discuss a series of questions given to them. |
| Element symbol and atomic size. Element symbol and shells. Element symbol and valencies. | About 13–14 | Students work in small groups to discuss a series of questions given to them. |
| Element symbol and first ionisation energies. Element symbol and electronegativities. | About 16–17 | Students work in small groups to discuss a series of questions given to them. |

At the same time, Periodic Tables are often presented in texts with numerous features of the elements shown in each box. This is simply beyond the capacity of the working memories of '*novice*' learners. This may be economic in space, but it is largely unintelligible to the '*novice*' learner. One way around this is to present Periodic Tables with only the element name (or symbol) and just one feature shown or present the data in the form of a graph showing the trend. The learners then work in groups to discuss various questions that draw out possible trends at a descriptive level. This can lead naturally on to interpretation of why such trends exist. Numerous aspects can be covered easily (Table 5.2).

Examples include discussing the dates of discovery of the elements, based on the work of Al-Osaimi (2013, p. 407). This was used with 12–15 year old students and was used originally in Arabic but is shown in English here in full to illustrate a simple idea:

> It takes time to keep writing down the names of the elements. Many years ago, each element was given a **SYMBOL**. Unfortunately, all the symbols are not in Arabic but use Western languages. This is because all the elements were first discovered by scientists working in the West. This makes it much more difficult for those who speak Arabic, or Chinese, or many other languages!
>
> Here is the table using these symbols, showing the date when each element was first obtained pure (Figure 5.16).
>
> (1) The dates of discovery for some elements are not given. Can you suggest a reason why they are not shown?
>
> ..........................................................................................
>
> (2) Why were metals like copper, silver, gold, tin and lead discovered so early?
>
> ..........................................................................................

**Dates of Discovery of the Elements**

| | | | | | | | | | | | | | | | | | |
|---|---|---|---|---|---|---|---|---|---|---|---|---|---|---|---|---|---|
| $_{1}$H 1776 | | | | | | | | | | | | | | | | | $_{2}$He 1895 |
| $_{3}$Li 1817 | $_{4}$Be 1828 | | | | | | | | | | | $_{5}$B 1808 | $_{6}$C | $_{7}$N 1772 | $_{8}$O 1776 | $_{9}$F 1886 | $_{10}$Ne 1898 |
| $_{11}$Na 1807 | $_{12}$Mg 1808 | | | | | | | | | | | $_{13}$Al 1827 | $_{14}$Si 1824 | $_{15}$P 1669 | $_{16}$S | $_{17}$Cl 1774 | $_{18}$Ar 1894 |
| $_{19}$K 1807 | $_{20}$Ca 1808 | $_{21}$Sr 1876 | $_{22}$Ti 1791 | $_{23}$V 1801 | $_{24}$Cr 1797 | $_{25}$Mn 1774 | $_{26}$Fe | $_{27}$Co 1735 | $_{28}$Ni 1751 | $_{29}$Cu | $_{30}$Zn 1746 | $_{31}$Ga 1875 | $_{32}$Ge 1886 | $_{33}$As 1649 | $_{34}$Se 1817 | $_{35}$Br 1826 | $_{36}$Kr 1898 |
| $_{37}$Rb 1861 | $_{38}$Sr 1808 | $_{39}$Y 1828 | $_{40}$Zr 1824 | $_{41}$Nb 1801 | $_{42}$Mo 1782 | $_{43}$Te 1937 | $_{44}$Ru 1844 | $_{45}$Rh 1803 | $_{46}$Pd 1803 | $_{47}$Ag | $_{48}$Cd 1817 | $_{49}$In 1863 | $_{50}$Sn | $_{51}$Sb 1620 | $_{52}$Te 1782 | $_{53}$I 1811 | $_{54}$Xe 1898 |
| $_{55}$Cs 1860 | $_{56}$Ba 1808 | $_{57}$La 1839 | $_{72}$Hf 1923 | $_{73}$Ta 1903 | $_{74}$W 1783 | $_{75}$Re 1925 | $_{76}$Os 1803 | $_{77}$Ir 1803 | $_{78}$Pt 1735 | $_{79}$Au | $_{80}$Hg | $_{81}$Tl 1861 | $_{82}$Pb | $_{83}$Bi 1753 | $_{84}$Po 1898 | $_{85}$At 1940 | $_{86}$Rn 1900 |
| $_{87}$Fr 1939 | $_{88}$Ra 1898 | $_{89}$Ac 1889 | | | | | | | | | | | | | | | |

**Figure 5.16** Dates of Discovery of the Elements.

(3) Many of the metals in columns I and II were discovered near the start of the 19th century. These are common metals. Why were they discovered so late in the world's history? What allowed them to be discovered at this time?

..........................................................................................

(4) Look at the last column (the noble gases). Where do most of these gases occur in the world? Why do you think they were they all discovered at a similar time?

..........................................................................................

(5) Nitrogen, oxygen and chlorine were all discovered about the same time. Can you suggest a reason why? Why was fluorine discovered later?

..........................................................................................

..........................................................................................

The questions allow the freedom to think widely and share ideas. At the end, the teacher can draw things together and all kinds of useful chemistry insights arise simply by looking at patterns in the dates. In the version used by Al-Osaimi, he provided a set of '*answers*' for the teacher to guide final discussion. Macro chemistry therefore offers considerable scope for useful discussion.

> Many claims are made about the value of small group discussion but some of these are undermined by the research evidence. In the next chapter, we shall look at the potential of small group discussions in the context of chemistry and draw in the key evidence from wider studies.

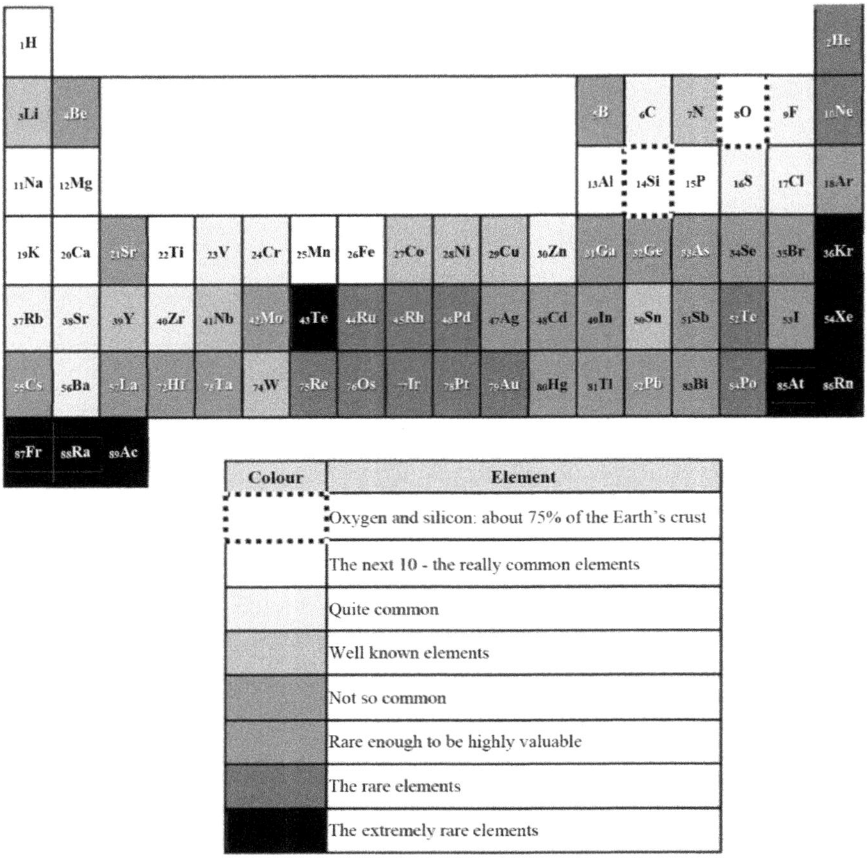


**Figure 5.17** Element Occurrence.

It is possible to generate all kinds of useful discussion by looking at how common the various elements are, common in the sense of being accessible humans by looking at the atmosphere, the sea and the land to a depth accessible to current mining technologies. The elements are placed in eight categories and coloured shading is really required to show the eight categories. However, the categories are shown here by various shadings to illustrate the approach (Hussein, 2006, pp. F5–F6) (Figure 5.17).

(1) Almost 75% of the Earth's crust is made up of the elements ***oxygen*** and ***silicon***. Where (air, sea, water or rocks?) might you find these elements in the earth's crust?

.......................................................................................

(2) Look at the elements shaded dark green (white in Figure 5.17). After oxygen and silicon, these are the top ten most common elements. Are there any surprises and can you explain why they are there?

Surprises:

..............................................................................

Reasons:

..............................................................................

(3) Many of the elements in columns I and II are common. Where would you find these elements (air, sea, water or rocks)?

..............................................................................

(4) Why are ***rare*** elements like silver, gold, platinum, helium and neon so well known?

..............................................................................

(5) Look at element 38 – Strontium. Use the internet or books to find as much as you can about this element. In what country was it discovered? After what was it named? Has it any uses?

..............................................................................

..............................................................................

The answer to these questions raises all kinds of potential discussion topics.

One of the sad things in so many chemistry curricula today is the way the coverage of the Periodic Table is often presented in such a dull and inaccessible way. The development of the table is one of humankind's greatest achievements. Its power to rationalise, to summarise and to reveal patterns and trends—all immense. Let us release the table to its full potential and use exciting ways to show school students its immense power. Johnstone has pioneered the way. Many others have taken it further. All we need to remember is the limitations of working memory capacity. With that warning, we now have a useful way to bring the macro to life with all kinds of insights from the sub-micro, sometimes bringing in the representational.

## 5.9 Key Messages

In order to communicate the ideas and understandings of chemistry to the next generation, we employ language, symbolisms, visual representations and models. Johnstone was acutely aware that, no matter how we communicate, the limitations of the working memory capacities of the learners control the extent of understanding. We need to use language carefully, especially when normal English words carry a very specific meaning in chemistry.

In everything, Johnstone argued that we start with the tangible, the macro, what can be experienced with the senses. Once the macro is well-established with the learners, then we can move step-by-step to introduce the sub-micro to interpret and explain what is observed. Inevitably, we need to represent the world at the molecular level in some way. The systems of representation

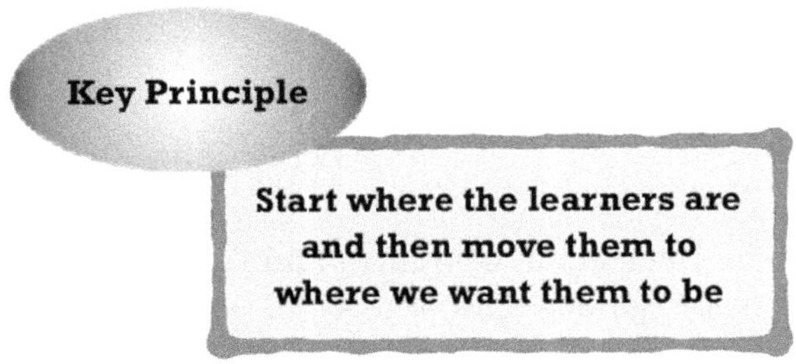


**Figure 5.18** Key educational principle.

we employ need explained and then introduced step-by-step. At each stage, time is needed for the learners to practise so that they become thoroughly familiar with the representation systems being used (Figure 5.18).

In so much curriculum planning and textbook writing, we are too strongly influenced by the logic of the subject to be taught. Long ago, Ausubel (1968) demonstrated, from his research, that knowing what the learners already knew and understood was critical. Understanding (or '*meaningful learning*', using Ausubel's language) involves new ideas being linked correctly on to ideas already held in long-term memory. One central aspect of this in chemistry is to start with the macro with '*novice*' learners. However, the macro must be meaningful. That is where the applications-led curriculum is so useful (Reid, 1999, 2000).

It was the fashion, at one time, to take groups of school chemistry students around, say, a production unit in an oil refinery or some other industrial chemistry site. If we are not careful, it can end up as a meaningless study of large pipework! A good video can often be much more meaningful. This stresses again that we must start where the learners are, in terms of what they know and understand, their age and their social context.

Whether we use language, the symbolisms of chemical or mathematical equations, diagrams, graphs or physical models, two key factors must be considered:

(1) Whatever we do must be presented in a way that does not overload limited working memory capacity.
(2) The goal is the development of *mental* models of physical reality where the models fit the evidence and make some sense of it: the goal is correct understanding.

Our goal is to enable the learners to observe and describe the molecular work using their senses and then, step-by-step, to draw them into possible ways to interpret and represent that world. How we achieve that will depend on the topic being considered.

We are not trying to teach learners that there are three levels. We are not aiming to show them how to link levels. Instead, we are allowing them to be chemists, for the moment, and look at possible ways to make sense of what they see and experience. We then move on to develop ways to represent the understandings so that they can be shared. This is the world of chemistry and chemistry has been doing all this for hundreds of years.

Faced with a class of young school learners, Johnstone wanted them to experience a little of being enabled to look at their world through the eyes of a chemist and enter into the wonder of making some sense of their world from a molecular standpoint.

**Evidence**

One experienced chemistry schoolteacher used to tell his students, '*do not believe anything I tell you unless I give you the evidence*'.

He then went on to invite the learners to ask for the evidence if, at any time, they felt they have not been given it.

Lessons from then on were certainly exciting and often intellectually challenging for the teacher!

## References

Ali A. A. and Reid N., (2020), *Making Sense of Learning, A Research-based Approach*, Berlin: Springer (in press).

Al-Osaimi K. H., (2013), *The Development of Critical Thinking Skills in the Sciences*, PhD thesis, Dundee: University of Dundee, p. 376, http://discovery.dundee.ac.uk/portal/en/theses/the-development-of-critical-thinking-skillsin-the-sciences(f8eb450c-0914-4e1b-aa1c-65dd9d02cfba).html, Accessed 5 November 2019.

Ausubel D. P., (1968), *Educational Psychology: A Cognitive View*, New York: Holt, Rinehart and Winston.

Baddeley A. D., Eysenck M. W. and Anderson M. C., (2015), *Memory*, 2nd edn, Glasgow: Psychology Press.

Ball P., (2020), Lies to Children, *Chem. World*, **17**(5), 17.

Bruner J. S., (1960), *The Process of Education*, Cambridge MA: Harvard College.

Bucat B. and Mocerino M., (2009), Learning at the Sub-micro Level: Structural Representations, in Gilbert J. K. and Treagast D. (ed.), *Multiple Representations in Chemical Education*, Dordrecht: Springer.

Cassels J. R. T. and Johnstone A. H., (1977a), *Language in Chemistry - Research for the Classroom and Beyond*, London: The Chemical Society, vol. 48.

Cassels J. R. T. and Johnstone A. H., (1977b), Language in Chemistry, *Scottish Centre for Mathematics Science and Technical Education*, **11**, 11.

Cassels J. R. T. and Johnstone A. H., (1978a), *Words in Chemistry*, London: The Chemical Society.

Cassels J. R. T. and Johnstone A. H., (1978b), What's in a Word, *New. Sci.*, **78**, 432.

Cassels J. R. T. and Johnstone A. H., (1980), *Understanding of Non-Technical Words in Science*, London: The Chemical Society.
Cassels J. R. T. and Johnstone A. H., (1983), Meaning of Words and the Teaching of Chemistry, *Educ. Chem.*, **20**(1), 10–11.
Cassels J. R. T. and Johnstone A. H., (1984), The Effect of Language on Student Performance on Multiple Choice Tests in Chemistry, *J. Chem. Educ.*, **61**, 613–615.
Cassels J. R. T. and Johnstone A. H., (1985), *Words that Matter in Science*, London: The Royal Society of Chemistry.
Davidowitz B. and Chittleborough G., (2009), Linking the Macroscopic and Sub-microscopic Levels: Diagrams, in Gilbert J. K. and Treagast D. (ed.), *Multiple Representations in Chemical Education*, Dordrecht: Springer.
Hindal H., (2007), *Cognitive characteristics of students in middle schools in State of Kuwait, with emphasis on high achievement*, PhD thesis, Glasgow: University of Glasgow, http://theses.gla.ac.uk/5717/, Accessed 9 November 2019.
Hindal H., (2014), *Cognitive Characteristics of High Achievers*, Germany: Lambert Academic Publishing.
Hindal H., Reid N. and Badgaish M., (2009), Working Memory, Performance and Learner Characteristics, *Res. Sci. Technol. Educ.*, **27**(2), 187–204.
Hindal H., Reid N. and Whitehead R., (2013), A Fresh look at High Ability, *Int. J. Instr.*, **6**(1), 59–76.
Hussein F. A. A., (2006), *Exploring attitudes and difficulties in school chemistry in the Emirates*, PhD thesis, Glasgow: University of Glasgow, http://theses.gla.ac.uk/6581/, Accessed 1 November 2019.
Hussein F. and Reid N., (2009), Working Memory and Difficulties in School Chemistry, *Res. Sci. Technol. Educ.*, **27**(2), 161–186.
Johnstone A. H., (1971), Conceptual Problems Associated with d-Orbital Models, *J. Chem. Educ.*, **48**(1), 74.
Johnstone A. H., (2000), Teaching of Chemistry – Logical or Psychological? *Chem. Educ.: Res. Pract. Eur.*, **1**(1), 9–15.
Johnstone A. H., (2010), You Can't Get There from Here, *J. Chem. Educ.*, **87**(1), 22–29.
Johnstone A. H. and Otis K., (2006), Concept mapping in problem based learning: a cautionary tale, *Chem. Educ. Res. Pract.*, 7(2), 84–95.
Johnstone A. H. and Morrison T. I., (1964), *Chemistry Takes Shape*, Book 1, London: Heinemann.
Johnstone A. H. and Morrison T. I., (1965), *Chemistry Takes Shape*, Book 2, London: Heinemann.
Johnstone A. H. and Morrison T. I., (1966), *Chemistry Takes Shape*, Book 3, London: Heinemann.
Johnstone A. H. and Morrison T. I., (1967), *Chemistry Takes Shape*, Book 4, London: Heinemann.
Johnstone A. H. and Morrison T. I., (1969), *Chemistry Takes Shape*, Book 5, London: Heinemann.

Johnstone A. H., Morrison T. I. and Reid N., (1981), *Chemistry About Us*, London: Heinemann, (reprinted 1982, 1986, 1988).

Miller G. A., (1956), The magical number seven, plus or minus two: some limits on our capacity for processing information, *Psychol. Rev.*, **63**, 81–87.

Reid N., (1999), Towards an Application-Led Curriculum, *Staff Educ. Dev. Int.*, **3**(1), 71–84.

Reid N., (2000), The Presentation of Chemistry: Logically Driven or Applications Led? *Chem. Educ. Res. Pract.*, **1**(3), 381–392.

Osgood C. E., Suci C. J. and Tannenbaum P. H., (1957), *The Measurement of Meaning*, Urbana, IL: University of Illinois Press.

Silverman L., (1989), The Visual-Spatial Learner, *Pre. Sch. Fail.*, **34**, 15–20.

Silverman L., (2002), *Upside-Down Brilliance: The Visual-Spatial Learner*, Denver, CO: DeLeon Publishing.

CHAPTER 6

# *Releasing Understanding: Three Broad Approaches to Teaching*

## 6.1 Looking Back

In Chapter 3, we saw that Johnstone's central argument was that we start with the macro chemistry, which can be quite rigorous in nature. When that is well-established, we then carefully build on the sub-micro and representational. Chapters 4 and 5 then moved on to look at ways by which we can add on the sub-micro and representational. Chemistry is communicated by means of language and visual representations. Care needs to be taken over languages, especially with those ordinary English words which chemistry employs with specialised meanings.

The visual–spatial is very important for many learners who tend to see things in terms of relationships between ideas and less by the logical sequences of language and number. Diagrams can help in bringing several ideas together. However, we need to be careful that the diagrams are not making excessive demands on limited working memory space. We saw the important principle that, in developing any kind of visual representation or physical model, we need to not only know the chemistry involved but, perhaps even more importantly, we need to know our learners. This requires that we understand how they see things, what their prior understandings are as well as the very real danger of overloading cognitive capacity.

Johnstone had a great flair in developing physical models that accurately and helpfully assisted learners in achieving understanding. He was also acutely aware that we need to avoid inappropriate models which can so often lead to misconceptions or may, indeed, require later '*unteaching*'. His use of

Advances in Chemistry Education Series No. 6
The Johnstone Triangle: The Key to Understanding Chemistry
By Norman Reid

Published by the Royal Society of Chemistry, www.rsc.org

modelling balloons has been found to be insightful as well as fun for the learners. They make the sub-micro very tangible. However, there was the added bonus. Implicit in the balloon model is the probabilistic nature of electron behaviour which lays the foundation for later quantum concepts.

He developed numerous practical ways to illustrate chemical concepts. His simple idea of tubes containing equi-molar quantities of solids (*e.g.* sodium halides) or liquids (*e.g.* alkanols) was a clever way to make size meaningful at the molecular level. He understood the power of the Periodic Table, employing it by looking at only one feature at a time and then developing graphs of atomic number against that feature. This brought trends among the elements to life. In a sense, he was making the sub-micro tangible or macro in nature.

## 6.2 Teaching for Understanding

In this chapter, the aim is to look at the implications from the Johnstone triangle and its underpinning message related to the key role of working memory for aspects of chemistry education, specially at school level but also as it applies in higher education. We shall start by considering the three main aspects for all teaching today (Table 6.1).

The division into the three areas is somewhat artificial for much lab work, for example can involve pairs while, in small groups working collaboratively together can involve tasks to be undertaken. Equally, a school lesson built around a didactic approach can often involve dialogue, debate and discussion. However, we shall look at these three areas to explore the implications of the working memory findings.

First, we need to undermine some of the myths that exists about all three. The first myth goes back to Confucius (6th century BC) where it is claimed that he said:

*I hear and I forget. I see and I remember. I do and I understand.*

We need to challenge this for it is often simply not true. There is now a totally consistent pattern of evidence that shows that understanding is *controlled* by the limitations of working memory capacity (Johnstone and El-banna, 1986, 1989). Johnstone's triangle developed from that key finding. As Johnstone found, it is possible to be very active in doing things in, say, the laboratory and come away having understood almost nothing (Johnstone and Wham, 1979, 1982).

**Table 6.1** Three aspects of teaching.

| | |
|---|---|
| Didactic | Teaching as telling. |
| Practical Work | Teaching as doing. |
| Group Work | Teaching as teamwork. |

The central role of limited cognitive capacity explains how understanding can be developed successfully. Long ago, David Ausubel (1918–2008) observed that the extent of teacher-centredness was *unrelated* to the extent of learner understanding (Ausubel, 1968). This can be presented as a simple diagram (Figure 6.1). The rote-meaningful axis is orthogonal to the reception-discovery axes. There is no correlation at all between the two dimensions.

There is a fashion today to suggest that placing learners in small groups with tasks to undertake or problems to solve will generate enhanced understanding. What the research consistently reveals is that sometimes it does, sometimes it does not. As Ausubel found, the extent of learner involvement *in this sense* is irrelevant. The key for enhanced understanding is *mental* involvement. That is controlled by the working memory. The working memory's capacity *controls* what in possible.

In a much more recent review of the research evidence, three eminent cognitive psychologists bring the evidence together to present a paper that probably should be mandatory reading for all involved in educational research and all involved in teaching at all levels. The paper is founded on clear evidence but is *uncomfortable* reading (Kirschner *et al.*, 2006). The subtitle for the paper is: '*An Analysis of the Failure of Constructivist, Discovery, Problem-based, Experiential, and Inquiry-Based Learning*'.

The authors are *not* arguing that constructivism is an incorrect description. They are *not* arguing that there is no value in discovery, problem-based, experiential, and inquiry-based learning. *They do show that none of these holds the key to understanding.* They stress that the working memory

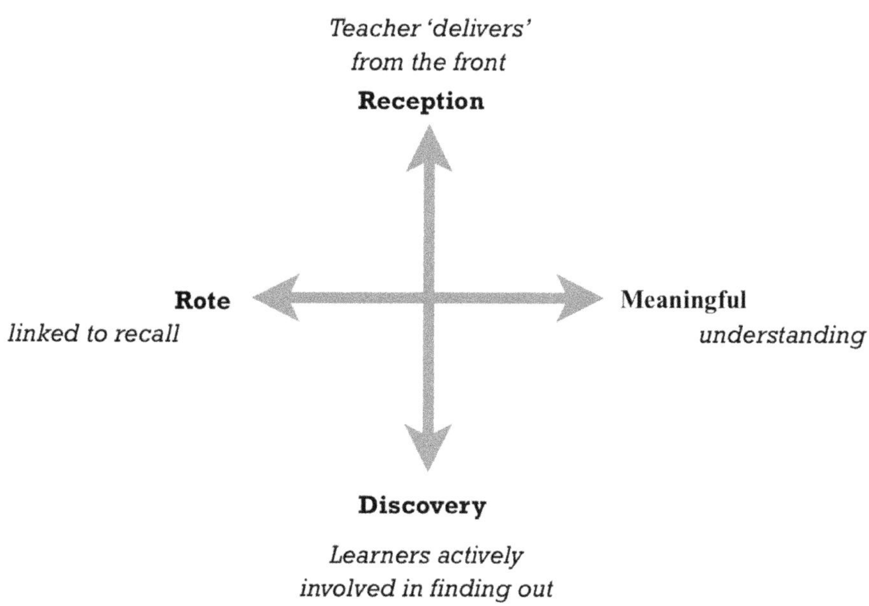


**Figure 6.1** Ausubel's finding.

holds the key and that cognitive overload will undermine possibilities of understanding.

We might be able to summarise this:

> Teaching method does **not** hold the key to learner understanding.
> Employing **any** method within the working memory capacity of the learners holds the key.

There is a simple set of principles related to teaching methods, all based on extensive evidence:

- Extent of understanding gained does not depend on the teaching method employed.
- Extent of understanding gained does depend on the method being used within working memory capacity limitations.
- The teaching method employed is very important for the development of skills and attitudes.

The question then arises what is the value in employing group work or practical work (like working in the laboratory) if there is no guarantee that understanding will be enhanced by these approaches. The answer is that research shows that there are other enormous advantages in these approaches. We shall return to these in a moment. However, in the context of chemistry, the Johnstone triangle comes into play yet again: start with macro and then introduce the other two levels step by step, with caution.

Our goal is to give to our students some insights into the way the molecular world works, interpreted by means of explanations at the sub-micro level. That implies the use of various forms of representation. Laboratory work can bring immense benefits. Group work can be very useful but there are two key conditions:

(a) The limited capacity of working memory must be taken into account.
(b) Didactic approaches, the use of group work, the place of laboratory work—all must be employed for their specific strengths and these are now known from extensive research evidence.

In education, today we have student-centred learning, problem-based learning, inquiry-based learning and discovery learning. There are many who think that, '*learning needs to be based on the needs and interests of the student and is therefore learner-centred*' while others argue that, '*learning needs to be explicit and rigorous and is therefore teacher-centred*' (Willis *et al.*, 2015, p. 19). The authors note the weakness of these viewpoints. Both are merely expressions of opinion. Indeed, there is today considerable loose thinking in the entire area of teaching methods. Firstly, we need clear shared descriptions of what the phrases mean so that we can start with shared

**Table 6.2** Key descriptions.

| Approach | Description | Key Reference |
|---|---|---|
| *Student-centred learning* | Methods of teaching that shift the focus of instruction from the teacher to the student. | Ali and Reid, 2020, chapter 16 |
| *Problem-based learning* | Where students, or groups of students, focus on a problem to be addressed and they have to gather the resources, insights and understandings in order to reach some kind of '*solution*'. | Mackenzie *et al.*, 2003 |
| *Inquiry-based learning* | This describes approaches to learning that are *driven* by a process of enquiry which actively involves students in discussion, questioning, and investigation. | Kahn and O'Rourke, 2004 |
| *Discovery learning* | The students gain practice in discovering things for themselves, the aim being to gain deeper understanding and greater skills in using understandings to solve problems. | Bruner, 1961 |

understandings. Table 6.2 offers a very brief summary, with one key reference provided.

There are overlaps between the various categories, but this gives us a shared starting language. One of the features that often occurs in all four areas is that students work together in small groups to achieve some task or to solve some problem.

There are extensive examples of problem-based activities at school level (Johnstone *et al.*, 1981a, 1981b; Yang, 2000; Al-Shibli, 2003; Chu and Reid, 2012). At university level, the approach is often adopted in the design of the entire course in medicine (Mackenzie *et al.*, 2003) while aspects of problem-based learning can be seen in numerous discipline areas. Research has shown great gains in the development of attitudes and skills arising from these approaches (Johnstone *et al.*, 1981a, 1981b; Johnstone and Reid, 1981; Reid, 1982). In inquiry-based learning (also known as enquiry-based learning), the entire process of learning is *driven* by the process of inquiry. The study by Lucas and Rowley (2011) showed gains in numerous generic skills. George-Williams *et al.* (2018) describe a project where a university laboratory course was re-designed reflecting aspects of context-based, inquiry-based and problem-based learning. While there were challenges, the evidence shows the development of skills which enable the students to be prepared better for life beyond the university.

With discovery learning, Bruner (1916–2016) assumed it was the best way allow learners to discover facts and relationships for themselves. This assumption was somewhat misplaced (van Merriënboer and Kirschner, 2013; De Bruyckere *et al.*, 2015; Sweller *et al.*, 2016). The learner does need some guidance. However, guided discovery learning seems to hold considerable potential. The critical importance of the teacher in the development of understanding was demonstrated in the meta-study published by Hattie (2009).

When it comes to understanding, many authors (Sweller *et al.*, 1998; Renkl and Atkinson, 2003; Mayer, 2004; Sweller *et al.*, 2016) all show that the teaching method is not the critical factor. The limitations of working memory capacity control extent of understanding. If any of these approaches is employed in such a way that overload is minimised, the approach may bring benefit. If overload is not taken into account, the approaches will not be effective. Indeed, there can be quite subtle disadvantages in employing these approaches, without adequate teacher support.

Therefore, Hattie and Yates (2013, p. 78) note that

> *... several studies have found that low ability students will prefer discovery learning lessons to direct-instruction-based lessons but learn less from them. Under conditions of low guidance, the knowledge gap between low and high ability students tend to increase. The lack of direct guidance has greater damaging effects on learning in low ability students....*

All these approaches have great value when seen in terms of the development of attitudes and skills. There is no evidence that they bring any universal benefits in terms of gaining information and enhancing understanding. However, they can be useful in allowing learners to *apply* previous understandings. At the same time, such approaches can be very costly in terms of time for teacher preparation (Willis *et al.*, 2015, p. 37).

If we look at all the various forms of student-centred learning approaches, the overall research findings indicate two central findings:

(1) There is no evidence that shows that such approaches bring any universal advantages in terms of gaining knowledge and understanding.
(2) There is good evidence of gains in other areas including the development of attitudes and skills.

## 6.3 Teaching as Telling

One of the sad features of much thought in wider society is that schools and universities exists to pass on information from teacher to learners who, in varying degrees, show they have memorised that information by gaining good examination marks. This is reflected in the way the media and politicians look at national examination outcomes. It is reflected in the numerous TV quiz programmes where success usually comes from speed of recall of memorised information.

Of course, the teacher (as an expert) has a task of communicating knowledge, understanding and experience to the learners (novices). With large numbers, this must involve an element of didactic presentation. However, the real goal relates to *understanding* and we can define understanding in terms of the ability to apply what we know with a good prospect of success. The didactic approach may be efficient in terms of the transfer of knowledge from '*expert*' to '*novice*' from a perspective of time and cost. However, we must question if this is meaningful education.

It is here that the work of Johnstone is so important in chemistry education. The central place of working memory limitations has been established in the context of didactic teaching (such as lecturing) for a very long time (Berliner, 1969; Anderson, 1980; Kiera, 1982). Given the dates on these studies and the wide range of publications relating to the controlling influence of working memory and specifically, the numerous papers written by Johnstone on the theme, it is utterly surprising that the findings are largely ignored in much school chemistry education.

**A Puzzling Conundrum**

The findings related to working memory and its control over understanding have been known for about 50 years.

The publications of Johnstone specifically relating limited working memory capacity to understanding in chemistry have been available for over 30 years.

Why is this so rarely taught in teacher education and why is it so often ignored in curriculum construction, assessment and textbook writing?

What Johnstone did was to show that difficulties in understanding can be interpreted in terms of working memory capacity and potential overload. This then led on to the triangle model when he demonstrated that the working memory of the '*novice*' learner simply cannot cope with the three levels, *at the same time*. So often, the triangle model is acknowledged and, indeed, welcomed. However, very often, the meaning of the model is not being understood or applied.

The truly brilliant teacher possesses such empathy for the learners that material can be presented in such a way that it is accessible. Accessible simply means that the learners are able to bring together the key ideas in working memory, without overload, so that understanding is gained. Such a teacher or lecturer is continually developing insights, challenging thought, dialoguing literally or imaginatively with the minds of the learners.

Nonetheless, the didactic presentation has considerable limitations. Developing wider skills is not so easy: skills like critical thinking, skills of communication, cooperation, exploring options, engaging in open-ended problem solving, for example. This is where group-work tasks and laboratory experiences come into their own. However, they must be employed for their strengths.

At university level, some lecturers use their lectures (often giving fewer) to outline the landmarks of an area of chemistry. The students are then required, using online resources as well as textbooks, to '*work up*' the lecture topic and then to generate their own notes that reflect their own understandings. However, students need guidance and support to develop this approach. Indeed, they need to know at the outset that this is the intended approach!

Learners do not find it easy to know what is of central importance. The lecturer can assist. It is important to flag up central ideas, key items of information, key principles and central insights. This encourages student mental engagement with the material. However, the tutorials linked to such a lecture course need to be adjusted to allow students to interact their understandings with those of other students and the tutor.

In Chapter 2, we discussed the extensive findings related to field dependency. These emphasise how important it is, at both school and university levels, for the teacher to highlight the landmarks, showing what is central, what is peripheral. Field dependency is the skill in being able to focus in on what is essential for the task in hand, leaving aside what is not important for achieving that task.

Chemistry students need encouragement to look at chemical situations and develop the skills of focussing in on what is essential, setting aside what may be important but not important *for the task in hand*. It has been shown that this skill is one key way by which we avoid cluttering our working memories with inessential information (Johnstone and Al-Naeme, 1991). The pre-lecture idea that Johnstone developed is an illustration of this: it sets out the landmarks by linking back to past understandings held by the learners, ensuring that the important ones are bought to the forefront of their thinking, This allows them to make much more sense of new material simply because the overloading of working memories is less likely.

There are other features of didactic teaching that are important. We need to take our students into our confidence by explaining how we will teach and what is needed from them to gain understanding. At school level, many chemistry teachers have adopted a role where they see their task is to *work with* the learners so that, together, the learners are able to perform well, examinations being the inevitable, but potentially scary, experience that faces them all.

One school chemistry teacher was heard to explain to his students, near the end of their two years course when they faced their first national examination, that an examination was simply '*a game*', a game where the examiners made up the rules but did not share them. Our task was to work them out. The same teacher had found a simple way to enhance multiple choice performance by about 15%, simply by techniques based on insights gained by memory research undertaken by psychologists. The approach was largely based on the way the working memory is involved in '*cueing*' as it searches long-term memory for anything familiar. Table 6.3 outlines the kind of way that was being taught (the examination paper at that involved 40 multiple-choice questions).

In fact, this approach has advantages when applied to any examination paper. Reading it all through before lifting a pen starts the working memory in its search for pathways to potential answers as held in long-term memory.

The observation that examination success can be enhanced in this way by various strategies raises interesting questions. To what extent is examination success dependent on strategy as well as various aspects of human

**Table 6.3** Multiple-choice questions.[a]

| | |
|---|---|
| Read the entire multiple-choice paper through, answering nothing. | May take several minutes but write nothing. |
| Start at the beginning and answer only those questions where you are 100% sure. | Do not worry of this is only a handful of questions. |
| Start at the beginning again and now answer those questions where you are reasonably confident. | You may be able to answer about a third to one half. |
| Start at the beginning again and now answer the questions where you have some idea or recognition. | Another small group of questions here. |
| Start at the beginning again and now answer the questions where there is some vague recognition. | A few more. |
| Start at the beginning again and make intelligent guesses of the remainder. | The final few. |

[a]Never go back and check any answer you made previously.

psychology (like confidence, ability to keep calm) and not on the candidate's understanding of the subject? Our faith in the reliability of formal end-of-course assessment (set externally to the school) is touching but often misplaced. The evidence shows clearly that shorter multiple assessments during and at the end of the course are much more reliable (Harlen, 2005; Hayward and Spencer, 2006).

There are inevitable errors in any test or examination. For example,

- The marking can be inconsistent: no marker can be perfectly consistent.
- Circumstances: the candidate may not be able to perform to their best on a particular day.
- Chance: the particular questions suit a particular candidate.
- Validity: the paper does not test fairly against the curriculum or the curriculum goals.

If we employ two (or more) assessments on two (or more) different occasions, some of the errors cancel out. The accuracy of the assessment rises considerably. The issues have been reviewed (Harlen, 2005; Gardner *et al.*, 2010) while it is worth studying the practices of the International Baccalaureate (2010) which has developed useful and practical ways of assessment that do reflect the research evidence. However, we need to remember the detailed research that Johnstone directed relating to the use of multiple-choice questions in chemistry. This certainly casts considerable doubt on the reliability of this assessment format (Friel and Johnstone, 1979a, 1979b, 1988).

One other powerful way to enhance understanding is by means of the careful use of the visual. We have seen that the visual can often capture relationships and new ideas and bring them together so that the learner sees several ideas as one '*chunk*'. Modern multi-media can allow information to be presented bit by bit or a complex diagram to be built up step-by-step.

However, the learner cannot imitate this easily on paper. In everything, we must be cautious. In mathematics, the use of diagrams was found to be able to help or to hinder understanding (Alenezi, 2008). It all depends on working memory limitations. We are back to the messages underpinning the Johnstone triangle. Working memory holds the key.

## 6.4 Teaching as Sharing

One of the tragedies of much modern school education is that it encourages competition. There is competition between learners to gain the high marks. We applaud the most successful and award them prizes. We forget the others, some of whom leave their educational journey with a deeply ingrained sense of failure that continues on with them throughout life. Is this really an ethical way for education to develop?

There is the competition between schools and the production of league tables that can elevate or destroy the reputations of schools. Sadly, the evidence shows that the outcomes are largely illusional. They are unrelated to the real quality of what is going on in the schools. When this approach is adopted, teachers can end up disillusioned and parents are misled (Gordon and Reese, 1997). One visible outcome is the haemorrhage of teachers from the profession resulting in teacher shortages. This tends to be most acute in subjects like chemistry in that chemistry graduates often find other jobs more easily.

Then, there is competition between countries with international league tables. These generate endless '*reforms*', none of which can be shown to bring any benefit to learners. The universal evidence shows that such competitive approaches are counterproductive (Popham, 1999, 2010; Meyer and Schiller, 2013; Davis *et al.*, 2015). In life in wider society, by contrast, collaboration and cooperation are often the keys to success. In education cooperation and collaboration are often equated to '*cheating*' (Robinson, 2011). We have a strange paradox.

As soon as we invoke group work as a teaching approach, we are implying that there are small groups of learners *working together* on some theme or topic or seeking to solve some problem collaboratively. We have seen that there are no universal advantages in using group work in terms of understandings gained. The working memory limitations still hold the key. The messages from Johnstone's triangle apply here.

Nonetheless, many studies have shown the enormous value of group work in achieving other, often very important, goals (Reid, 1976, 1978, 1980; Johnstone *et al.*, 1981a, 1981b; Johnstone and Reid, 1981; Reid, 1982; Clarkeburn *et al.*, 2000; Reid and Yang, 2002; Hager, *et al.*, 2003; Lucas and Rowley, 2011; Al-Ahmadi and Reid, 2012; Al-Osaimi, *et al.*, 2015). Figure 6.2 summaries the key findings from many studies.

Group work involves extensive preparation in advance. It is much more than placing learners in small groups and asking them to discuss something. The best group work involves the learners with a task to complete or a

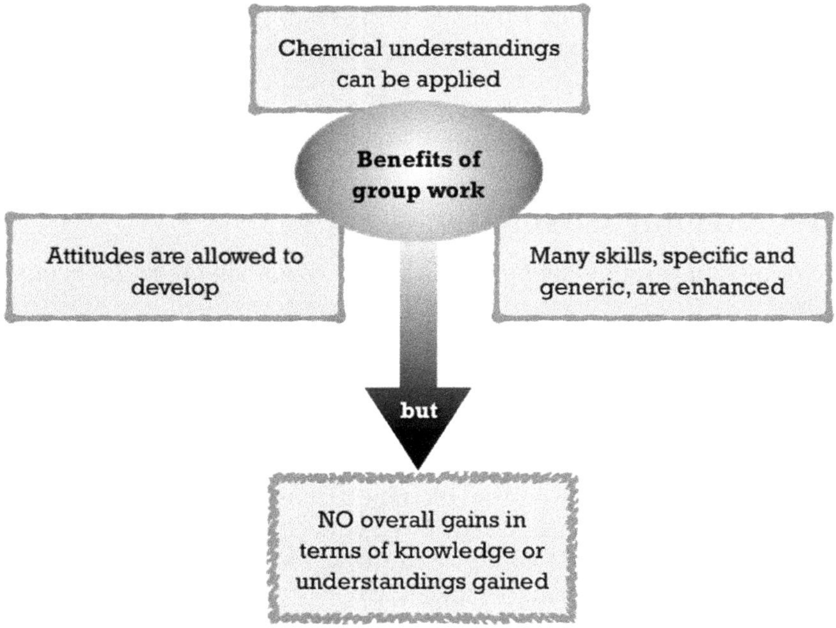


**Figure 6.2** Benefits arising from group work.

problem to solve. The best problems to solve are very different from the traditional kinds of problems we often think of when studying chemistry.

Most problems in chemistry are closed: the data are given; the method has been taught and the goal is specified. Johnstone (1993) analysed the entire nature of problem-solving and showed this was just one of eight types of problems. In varying degrees, the real problems that we face in life are much more open-ended. There is enormous scope for much more open problems and tasks in chemistry.

Johnstone (1993) observed that there are three variables associated with all problems:

*The data provided.*
*The method to be used.*
*The goal to be reached.*

By looking at the extremes where each variable is either fully known or unknown, this leads to eight problem types (Table 6.4).

Most problems usually encountered in textbooks and examination papers are essentially type 1 problems: the students are given information; they have been taught the method to be used and they are told what the goal is. In essence, they are '*exercises*'. Johnstone made it clear that the type of problem did not relate to difficulty and this finding was confirmed much later by Yang (Reid and Yang, 2002). The strength of this simple analysis is that it

**Table 6.4** Eight types of problems (Johnstone, 1993, Introduction).

| Type | Data | Methods | Goals/Outcomes |
|---|---|---|---|
| 1 | Given | Familiar | Given |
| 2 | Given | Unfamiliar | Given |
| 3 | Incomplete | Familiar | Given |
| 4 | Incomplete | Unfamiliar | Given |
| 5 | Given | Familiar | Open |
| 6 | Given | Unfamiliar | Open |
| 7 | Incomplete | Familiar | Open |
| 8 | Incomplete | Unfamiliar | Open |

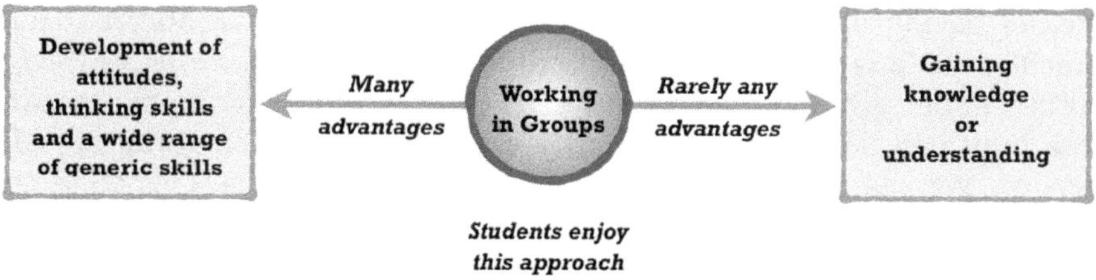


**Figure 6.3** Goals for group work.

gives a clear language to allow us to re-think the kinds of problems that are being used in chemistry education today. Therefore, it gives a practical basis by which group work problem-based tasks can be developed.

We also need to have clear goals. Group work has been employed extensively in chemistry at both school and university level and the key goals relate to the development of skills and attitudes (Johnstone *et al.*, 1981a, 1981b; Johnstone and Reid, 1981). In his meta-study, Hattie (2009) has demonstrated that there is *no overall advantage* in working in groups if our goals are understanding. This finding is too often ignored. However, the evidence equally shows that there can be enormous advantages in other areas (Figure 6.3). There is a bonus: learners enjoy this kind of work as well as developing important generic skills (Clarkeburn *et al.*, 2000).

At school level, it is important in setting up group work tasks for there to be a clear statement of what the learners are to do or to achieve. There will need to be some clear indications of how to approach the task and all the necessary information may need to be provided. There is some variation possible here. There are occasions where guidance and direction are given but the group is led to work out the detailed way to approach the task. At university level, there will be occasion where students are given much more freedom to find out the information they need (Lucas and Rowley, 2011). Examples will be given in a moment or two to illustrate what is possible and what has been done successfully.

For the moment, let us return to Johnstone's triangle. This stresses that '*novice*' learners are unable to work at the three levels *at the same time.* This will still apply in group work tasks. Therefore, we have to be careful in developing the materials that we do not end up overloading working memories and generating even more confusion. However, peer support can be an advantage: it is sometimes easier to ask a question of a fellow learner than admit to the teacher that we are lost! Equally, if structured carefully, there will be occasions when levels can be linked. Group work does provide opportunities for learners to engage mentally with chemistry, through debate and discussion. Group work does provide opportunities to apply chemical ideas and this may strengthen links between levels.

Overall, research shows the areas of benefit arising from group work tasks and these are brought together in Figure 6.4.

Assessment does pose major problems. We are so indoctrinated with the idea of having the goal of obtaining *'right answers'* that we forget that, in much of life, it is often a matter of finding *any* answer that works, or the *best answer* under given circumstances or even that there may be two answers that both will be helpful. There may even be times when compromise is needed. We can see this frequently in areas where the best reaction conditions arising from an analysis of the kinetics and thermodynamics of a reaction do not coincide. We have to compromise to choose the best acceptable conditions for the process. Indeed, there may be economic factors that may outweigh simple chemical analyses.

In such open-ended group work tasks, very often what decision is taken is not so important. It is how we arrive at the decision. This may seem to be a new way of thinking (Figure 6.5).

This gives us the basis for assessment. Attempting to assess students after completing open-ended teaching units will usually reward the wrong outcomes. Research (working with large samples of students) shows what involvement with group-work tasks will achieve for learners: mainly in the areas of skills and attitudes. However, we have no way to assess this *at an individual level*. The more valid approach is to give the credit for participation. If students have undertaken the group work tasks (and, perhaps,

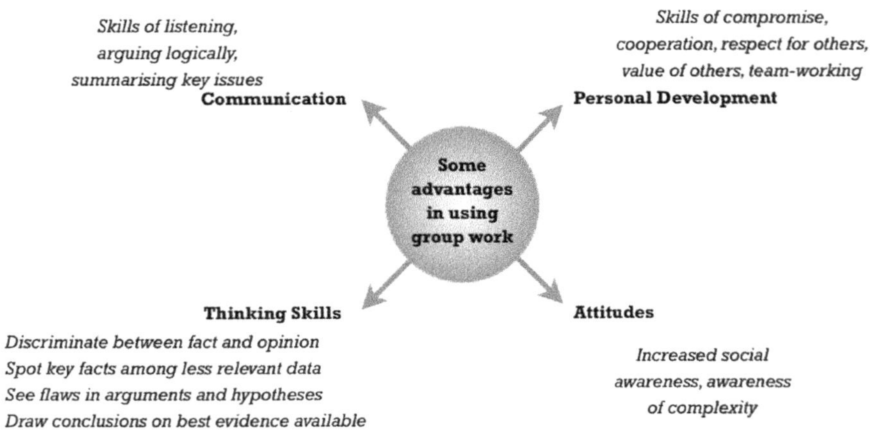


**Figure 6.4** Advantages in using group work in chemistry.

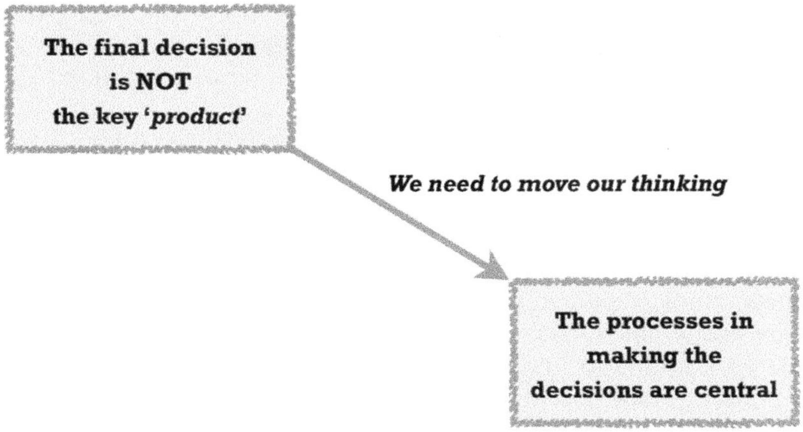


**Figure 6.5** Process, not product.

handed in a one or two paragraph report that merely provides the evidence of participation), credit can be awarded. This approach to assessment is known as '*duly performed*' and has been widely used. It is described in Ud-Din *et al.* (2016, pp. 88–93). It is also a powerful way to assess laboratory courses.

Let us look to see examples of group work activities from the literature. Working with Johnstone, Percival pioneered the group work approach with university chemistry students (Johnstone and Percival, 1978; Johnstone *et al.*, 1981a, 1981b). In one unit he developed, senior undergraduates were given some references relating to the development of polywater. Working in small teams, they followed these up and summarised their conclusions. They were then given some references that seemed to raise doubts that polywater existed. After following this up and forming their conclusions, they were given some references that totally undermined the concept. In groups, they reported the story overall. The goal of this group work exercise was to give the students an insight into the way chemistry, as a science, worked: experimental evidence was gathered, more evidence considered and the final understanding either supported or undermined. As a bonus, library skills were gained as well as the skills in summarising research.

At school level, sets of open-ended problem-solving materials were developed (Reid, 1976, 1980, 1982) and a set of 11 were included later in a book (Johnstone *et al.*, 1981a, 1981b). In one of the early developments, the production of sulfuric acid was considered. At that time, it was a central theme in the school chemistry curriculum. A class was divided into three groups (or multiples of three). The task for each group was to plan the construction of an industrial production unit.

The first group was given the data for the year 1873, the second for 1913 and the third for 1973. These dates were chosen because they were times when the industry faced difficult decisions. Each student in each group was given a small booklet for the relevant year. Working as groups, they had to take a series of decisions: raw material, factory size, location, product quality, process to be adopted.

At the end, the groups reported in on their decisions and found, to their amazement, that different groups had taken different decisions. The immediate reaction of the learners was to ask who was right. They are then told that they might all be right! The '*best*' answer for 1873 may be completely inappropriate for 1913 or 1973. One of the interesting outcomes was the way such an approach allowed the students to see how chemistry developments take place and how such developments have major implications for the way industry works and its impact on wider society.

In thinking of the objectives set for this group work unit, the task provides opportunities for students to identify problems and devise means of solving them, to argue logically and criticise the arguments of others, to make and defend decisions. Students learn that compromise may be a part of real science decision making, and that factors beyond science strongly influence possible solutions to problems. There are also opportunities for students to participate in cooperative problem solving, to exercise empathy, and to present a case in public.

Evidence was gained about the value of a group work unit like the sulfuric acid one (Reid, 1980). At the end, they had understood how the acid was made and that saved teaching time. They grasped how industrial decisions might be taken and the way industry changed over time. They began to appreciate how market demands influence industrial decisions both relating to product quality and to factory size. They were observed to interact with each other, to argue and debate about chemistry and its implications in wider life. They had thoroughly enjoyed the experience.

The material was later modified for use with undergraduates. The modification brought in some aspects of the kinetics and thermodynamics of the process. It also posed problems for the students as they had to solve pollution issues with ever more stringent legislation controlling emissions of sulfur dioxide. This gave scope for creative thought in seeking to modify the process design to minimise emissions of sulfur dioxide.

Another unit was developed that could be used at both senior school level and junior undergraduate level. Students were faced with a situation where the main chemical industry in a small community was threatened with closure. Tighter pollution requirements imposed financially impossible demands on the industry. Different groups played the roles of company management, the workforce, and the local politicians. The goal was to negotiate some kind of compromise. Each of the three role play groups was given the data as would have been provided by an industrial consultant but the data were not identical for the three groups, reflecting reality.

In another study, the study of spectroscopy with first year university students was built around enquiry-based learning. The evidence showed increased confidence among the students in the area of spectroscopy along with the development of a wide range of skills which can bring benefits in terms of future employability, including skills of independent learning, skills related to collaboration and group-working, as well as problem solving skills and communication skills (Lucas and Rowley, 2011).

With school curricula often being overcrowded (university curricula can sometimes be the same), time is an important issue. There are group-work units to fit into a one-hour tutorial slot or a three-hour laboratory slot. For schools, most of the units were shorter, often occupying a 40-minute lesson. All were built to fit into a specified curriculum, therefore saving teaching time elsewhere. They were not designed to focus on knowledge or understanding (although the evidence suggests that they may have been as effective as traditional teaching). They were designed to allow useful skills and attitudes to develop and the evidence shows that this did happen.

Developing such group work materials is very time-consuming and it is, therefore, important that they can be used widely. In an interesting follow-up using the same approach, Yang (2000) developed a set of shorter open-ended problem-solving group work exercises for school use in her study of problem solving. Later, Almadani (2012) took the chemistry units developed by others, adapted them to suit his country's curriculum and then translated them into Arabic. He was studying quality in education and the need for much less formal teaching and more group work activities in his country. This suggests one possible way forward is to second a creative chemistry teacher to a location with good library facilities for a short period of time. The teaching units developed by this teacher could then be shared widely across schools.

It has been found that the size of the group is important, to give every participant opportunity to contribute while avoiding '*passengers*' who contribute very little. Therefore, group size and its leadership can be very important (Hager *et al.*, 2003). Sometimes, leaders are appointed by the teacher, more often it is left to the group. The effective leader is a student who carries respect, being neither over-dominant nor over-compliant. The goal is to encourage all to contribute.

In looking at group work in chemistry, there is enormous scope here, not to enhance knowledge or understanding, but rather to develop important skills and allow attitudes to develop as well. Such units are best open-ended but with clear goals set for the students. The time to develop them can be considerable. They need to be developed to fit the curriculum being followed and be designed to avoid working memory overload. They allow chemistry understandings to be applied to real-life contexts and, in a threatening atmosphere, they allow students to interact with each other in areas of chemical importance, perhaps allowing the insights from the sub-micro and representational to link with the macro chemistry, albeit in a simulated setting.

Looking back again at the Johnstone triangle, group working holds considerable potential in applying understandings in chemistry. Inevitably, the macro will interface with the sub-micro and the representational.

In Chapter 4, we looked at one teaching unit where the students were allowed to re-live the research that led to the use of halothane as a key anaesthetic. They started by looking at the outcomes from some chemical research (possibly the kind that might be described as '*blue-sky*') that explored key properties of various complex haloalkanes.

Working in groups, there was considerable use of representations and many calculations were required. The students were only aged 15 but they suddenly discovered the value of chemical calculations that so often appear abstract and pointless. They had the wonderful experience of being able to identify some of the key compounds that led to the transformation of modern surgery. One feature was that the students could begin see the way research in chemistry made a real contribution for the benefit of humans.

## 6.5 Teaching as Doing

In Chapter 4, we looked at laboratory teaching, which was seen as one way by which the macro can be made real. Much of the research directed by Johnstone looked at laboratory work in a university setting. However, in some early work, he found that the new Scottish curriculum introduced into Scottish schools in 1962 was generating a very large amount of laboratory work for the learners (Johnstone and Wood, 1977). This was not being achieved by prescribing experiments nor by pressures from outside the schools. This happened because the laboratory work had a clear purpose that was integral to the way the curriculum had been constructed.

**School Laboratory Work**
This comes into its own when it has a clear purpose and is fully integrated into the way the curriculum is constructed.

The new curriculum in Scotland was based on *guided* discovery (Curriculum Papers 512, 1962). Learners were being encouraged to ask questions of the world around them and these questions were being answered by means of the experimental. Understanding arose because there was experimental evidence that needed interpreted. The laboratory work was totally integrated with all learning.

In most schools, all the teaching took place in laboratories. Teachers would discuss with students, work out experimental ways forward to answer questions and then, carefully guided by the teacher, would work out how to carry out the experiments. After the experimental, there was discussion about the possible interpretation and meaning of the outcomes. This generated insights and understandings. Sometimes, it generated the next set of questions to be addressed. The laboratories were equipped for chemistry, with resources available, in a flexible way. The teacher knew fairly precisely where the lesson was probably going but the students were working under the guidance of a teacher to plan and carry this out.

**Organising laboratory work**

At school level, there is a danger in having prescribed experiments, with the equipment all organised in advanced by skilled technicians.

The students have little opportunity to discuss, plan and to work with the teacher to develop the best way forward.

**The power of laboratory work can often be undermined by organisational efficiency.**

For school laboratory work to be at its best, there needs to be maximum flexibility. The experimental, the discussion and the formal teaching all need to be linked together. The evidence shows that the students liked the approach that Johnstone had introduced and opted into chemistry in later courses in increasing numbers. The way it was organised reflected the nature of chemistry as a science. Chemistry seeks to address questions of the physical world to explore how things work at the molecular level. The laboratory is very often where such questions can be addressed.

The way the course was conceptualised was strongly influenced by Johnstone whose ideas had underpinned the entire curriculum. Some countries pursued courses based on *discovery* (not *guided* discovery) but found that the students often could not cope. They needed strong guidance. The working memories of '*novice*' learners are rapidly overwhelmed in a subject like chemistry unless there is sensitive teacher support. Indeed, it is quite unreasonable to expect school students to make discoveries and develop insights that took centuries to emerge and to do it in a few lessons in a school laboratory.

Other countries, sensing the importance of laboratory work, laid out set experiments and provided laboratory accommodation separate from the teaching areas. This proved too inflexible. In many parts of the world, this still continues, and the evidence shows that the students do not enjoy laboratory work while they struggle to understand chemistry, the net outcome being that many opt out of chemistry as soon as they can (Hussein, 2006).

It all comes back to our aims for school chemistry. Get these right and this determines how we employ laboratory work. While laboratory work is very important for establishing the macro, Johnstone saw laboratory work as a wonderful opportunity to ask questions of the world around and, on the basis of experimental evidence, lead the learners into new insights and understandings. In that sense, laboratory work can be a great bridge between the macro and the other two levels in the Johnstone triangle (Tsaparlis, 2009).

Based on the evidence, five broad goals were suggested for school laboratory work in Chapter 4 (Table 6.5).

If we organise school laboratory work around periodic visits to a laboratory to watch or carry out some set experiments, we destroy the value of laboratory work. If we formalise laboratory work around the completion of set

**Table 6.5** Goals for school chemistry laboratory work.

| | |
|---|---|
| Making the chemistry real. | The essence of the macro level. |
| Revealing the methods of the sciences. | The central part of education for all. |
| Developing questioning skills. | The way the sciences move forward. |
| Generating understanding. | The ultimate goal. |
| Enhancing personal skills. | The great educational bonus. |

experiments to be assessed by marking laboratory reports or by observing if learners can demonstrate specific laboratory skills, we distort the value of laboratory work. If we employ laboratory manuals as recipes to follow, then we do not allow the learners to discuss and plan experiments and we undermine totally the nature of laboratory work. If we employ laboratories to verify '*theory*', we reverse the true nature of chemistry as a science. If we set up laboratory experiences that make so many demands on limited working memory capacity, then there will be no room for thought: understanding will be the casualty.

Laboratories are places where students can see how materials behave, albeit in somewhat artificial conditions. They are places where we can ask questions of our remarkable world and see how it reacts when we do things to it. They are places which encourage questioning, critical thinking, dialogue and debate and the development of mental models that seek to rationalise and make sense of the molecular world.

Chemistry is based on evidence and, like all the sciences, the evidence is mostly experimental. This reflects the true nature of chemistry. It is not chance that chemistry, as a modern science emerged in the 16th–17th centuries in Europe, often led by clergymen. This was the time in history when humans in the West started to break from believing things because some authority figure said it was so. The clergymen saw no essential difference between their chemistry explorations and their spiritual journey: they were looking at aspects of God's world, as they saw it.

At the end of Chapter 5, one teacher was described who told his students that they were not to believe anything he said, unless he gave the evidence. On one occasion, this teacher was caught out. He happened to throw in a passing remark that the reactivity of titanium was rather like that of aluminium and that explained why it was so difficult to extract from rocks. One student then quietly observed: '*but you never gave us the evidence*'.

The teacher went into the backroom off the lab to see if there was any titanium metal. He found a very small bottle with some pellets in it. The class have just seen the way aluminium reacted with oxygen, the pyrex test-tube melting and bending. With help from the class, they set-up the same apparatus (Figure 6.6), but in a safety protected fume cupboard as the teacher had never seen the oxidation of titanium before and was unsure exactly what might happen. The class knew that it was new to the teacher and their question had generated the need for the experiment.

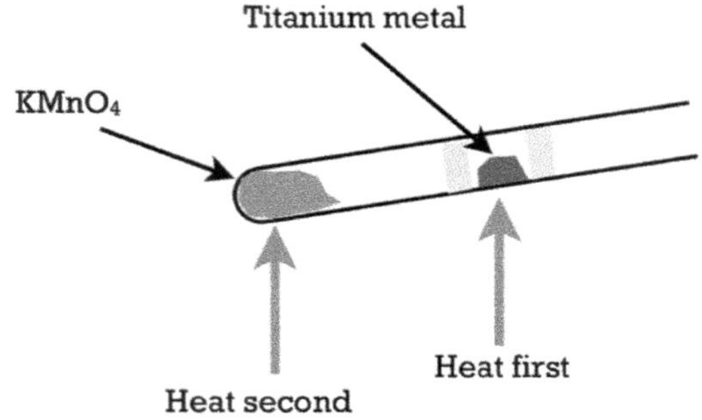


**Figure 6.6** Oxidation of titanium metal.

> Aluminium oxide could be electrolysed as a melt (adding cryolite made that possible at a temperature that was accessible). Titanium(IV) oxide was polar in nature and could not be electrolysed in a parallel way.

It was observed that the titanium oxidised in a similar way to aluminium, generating enormous amounts of energy. The test-tube melted and bent over. The class would never forget that experiment. Here is the correct use of laboratory work. We address questions of the macro world. Then we experiment with that world to see how it will react.

## 6.6 Key Messages

The goal of all chemistry teaching has to be centred on understanding: enabling the learners to make sense of the world at the molecular world. Simply because that world is inaccessible directly to human senses, there is an inevitable need for representations, often involving the sub-micro, in generating understanding. This immediately places the working memory under pressure as it seeks to cope with chemistry at three levels.

The extent of understanding is controlled by the limited and fixed nature of the working memory capacity. The extent of understanding is *not* controlled by the teaching method employed. Overall, in terms of knowledge and understanding gained, there is no advantage in employing group work activities or practical activities. However, group work has enormous advantages when seeking to achieve specific goals. Laboratory work can also achieve specific very valuable goals. These goals are very difficult to achieve easily in any other way. Therefore, group work and laboratory learning have a central role in chemistry education. However, both must take account of the limitations of working memory capacity and be planned in the light of the triangle model that summarises the nature of chemistry in the process of learning.

Based on research evidence, the overall pattern is illustrated in Figure 6.7.

All three approaches have their strengths and weaknesses. The place of group work is very important if we are seeking to give opportunities for

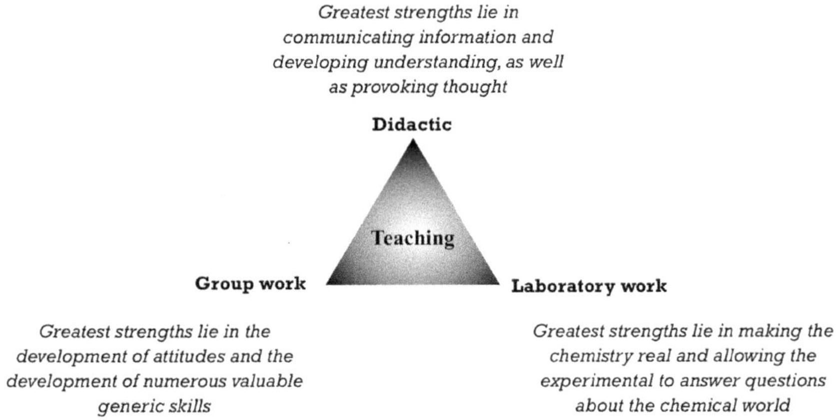


**Figure 6.7** Approaches to teaching.

learners to develop attitudes and ways of thinking as well as a wide range of very valuable generic skills. Such skills have an important place whether the individual learner follows a scientific career or moves into an area of life where chemistry seems to have little impact. Johnstone initiated the development of group work teaching materials for use at both school and university levels. Many of the early units were found to be powerfully effective in achieving the goals set for them. However, there is a need today for new units to be developed and older group work units brought up to date.

Laboratory work can provide a bridge between the macro and the other two levels. However, it depends on how the laboratory work is organised. In his earlier work, Johnstone developed powerful ways by which laboratory work could be employed to illustrate how chemistry addresses its questions to the world around us and then seeks to make deductions based on how that world reacts. This perspective can be seen as a central part of his early five textbooks (Johnstone and Morrison, 1964–1969) for school level and this approach offers a very useful template for the use of laboratory work at school level today.

In the next chapter, we shall move on to look at some of the very practical ways difficult concepts have been approached, seen from the perspective of the Johnstone triangle.

# References

Al-Ahmadi F. and Reid N., (2012), Scientific Thinking - Can it be Taught? *J. Sci. Educ.*, **13**(1), 18–23.

Al-Osaimi K. H., Reid N. and Rodrigues S., (2015), Critical Thinking - Can it be Developed? *J. Sci. Educ.*, **15**(2), 57–75.

Al-Shibli A. A. S., (2003), *A Study of Science Student Teachers Perceptions of Learning in the Education Colleges in the Sultanate of Oman*, PhD thesis, Glasgow: University of Glasgow.

Alenezi D., (2008), *A Study of Learning Mathematics Related to some Cognitive Factors and to Attitudes*, PhD thesis, Glasgow: University of Glasgow, http://theses.gla.ac.uk/333/, Accessed 7 November 2019.

Ali A. A. and Reid N., (2020), *Making Sense of Learning, A Research-based Approach*, Berlin: Springer.

Almadani K. A., (2012), *Quality Assurance Related to Secondary Education in the Kingdom of Bahrain*, PhD thesis, Dundee: University of Dundee, https://discovery.dundee.ac.uk/ws/portalfiles/portal/1630502/Almadani_phd_2012.pdf, Accessed 12 December, 2019.

Anderson J. R., (1980), *Cognitive Psychology and Its Applications*, San Francisco: W.H. Freeman.

Ausubel D. P., (1968), *Educational Psychology: A Cognitive View*, New York: Holt, Rinehart and Winston.

Berliner D. C., (1969), *Effects of Test-like Events and Note-taking on Learning from Lecture Instruction*, Washington: American Psychological Association.

Bruner J. S., (1961), The act of discovery, *Harv. Educ. Rev.*, **31**(1), 21–32.

Chu Y.-C. and Reid N., (2012), Genetics at school level: addressing the difficulties, *Res. Sci. Technol. Educ.*, **31**(1), 1–25.

Clarkeburn H., Beaumont E., Downie R. and Reid N., (2000), Teaching biology students transferable skills, *J. Biol. Educ.*, **34**(3), 133–137.

Curriculum Papers 512, (1962), *Alternative Chemistry for Ordinary and Higher Grade*, Dalkeith, Edinburgh: Scottish Examination Board.

Davis A., Winch C. and Lunn G., (2015), *Educational Assessment on Trial*, London: Bloomsbury Academic.

De Bruyckere P., Kirschner P. A. S. and Hulshof C. D., (2015), *Urban Myths about Learning and Education*, Amsterdam: Elsevier, pp. 1–16.

Friel S. and Johnstone A. H., (1979a), Does the Position of the Answer in a Multiple Choice Test Matter? *Educ. Chem.*, **16**(6), 175.

Friel S. and Johnstone A. H., (1979b), Seconds Thoughts in Multiple Choice Tests in Science, *J. Chem. Educ.*, **56**(5), 326.

Friel S. and Johnstone A. H., (1988), Making Test Scores Yield More Information, *Educ. Chem.*, **25**(2), 46–49.

Gardner J., Harlen W., Hayward L., Stobart G. and Montgomery M., (2010), *Developing Teacher Assessment*, Maidenhead: Open University Press.

George-Williams S. R., Soo J. T., Ziebell A. L., Thompson C. D. and Overton T. L., (2018), Inquiry and industry inspired laboratories: the impact on students' perceptions of skill development and engagements, *Chem. Educ. Res. Pract.*, **19**, 583–596.

Gordon S. and Reese M., (1997), High Stakes Testing; worth the price? *J. Sch. Leadership*, **7**, 345–368.

Hager P., Sleet R., Logan P. and Hooper M., (2003), Teaching critical thinking in undergraduate science courses, *Sci. Educ.*, **12**(3), 303–313.

Harlen W., (2005), Trusting teachers' judgement: research evidence of the reliability and validity of teachers' assessment used for summative purposes, *Res. Pap. Educ.*, **3**, 245–270.

Hattie J., (2009), *Visible Learning: A Synthesis of 800 Meta-Analyses Relating to Achievement*, New York: Routledge.

Hattie J. and Yates G. C., (2013), *Visible Learning and the Science of how We Learn*, New York: Routledge.

Hayward L. and Spencer E., (2006), There is no alternative ... to trusting teachers, in Sainsbury M., Harrison C. and Watts A. (ed.), *Assessing Reading - From Theories to Classrooms*, Berkshire: NFER, pp. 222–240, https://www.nfer.ac.uk/media/1610/asr01.pdf#page=232, Accessed 15 December 2019.

Hussein F. A. A., (2006), *Exploring Attitudes and Difficulties in School Chemistry in the Emirates*, PhD thesis, Glasgow: University of Glasgow, http://theses.gla.ac.uk/6581/, Accessed 1 November 2019.

International Baccaleaurate Assessment Practices, (2010), *Diploma Programme Assessment: Principles and Practice*, Geneva: International Baccalaureate Organization, [First published 2004].

Johnstone A. H., (1993), Preface, in Wood C. A., *Creative Problem Solving in Chemistry*, London: Royal Society of Chemistry.

Johnstone A. H. and Al-Naeme F. F., (1991), Room for scientific thought? *Int. J. Sci. Educ.*, **13**(2), 187–192.

Johnstone A. H. and El-Banna H., (1986), Capacities, Demands and Processes: a Predictive Model for Science Education, *Educ. Chem.*, **23**(3), 80–84.

Johnstone A. H. and El-Banna H., (1989), Understanding Learning Difficulties - A Predictive Research Model, *Stud. High. Educ.*, **14**(2), 159–168.

Johnstone A. H. and Morrison T. I., (1964–1969), *Chemistry Takes Shape*, vol. 1–5, London: Heinemann Educational.

Johnstone A. H. and Percival F., (1978), More Tools for the Job, *Chem. Br.*, **14**, 507.

Johnstone A. H. and Reid N., (1981), Towards a Model for Attitude Change, *Int. J. Sci. Educ.*, **3**(2), 205–212.

Johnstone A. H. and Wham A. J. B., (1979), A model for undergraduate practical work, *Educ. Chem.*, **16**(1), 16–17.

Johnstone A. H. and Wham A. J. B., (1982), Demands of Practical Work, *Educ. Chem.*, **19**(3), 71–73.

Johnstone A. H. and Wood C. A., (1977), Practical Work in its own Right, *Educ. Chem.*, **14**, 11.

Johnstone A. H., Morrison T. I. and Reid N., (1981a), *Chemistry About Us*, London: Heinemann, (reprinted 1982, 1986, 1988).

Johnstone A. H., Percival F. and Reid N., (1981b), Is Knowledge Enough? *Stud. High. Educ.*, **6**(1), 77–84.

Kahn P. and O'Rourke K., (2004), *Guide to Curriculum Design: Enquiry-based Learning*, Imaginative Curriculum Network, Manchester: Higher Education Academy, http://www.ceebl.manchester.ac.uk/resources/guides/kahn_2004.pdf, Accessed 7 November 2019.

Kiera K. A., (1982), *Note-taking and Review; the Research and Its Implications*, Logan: Utah State University.

Kirschner P. A., Sweller J. and Clark R. E., (2006), Why Minimal Guidance during Instruction Does not Work: An Analysis of the Failure of Constructivist, Discovery, Problem-based, Experiential, and Inquiry-Based Learning, *Educ. Psychol.*, **41**(2), 75–86.

Lucas T. and Rowley N. M., (2011), Enquiry-based learning: experiences of first year chemistry students learning spectroscopy, *Chem. Educ. Res. Pract.*, **12**(4), 478–486.

Mackenzie A. M., Johnstone A. H. and Brown R. I. F., (2003), Learning from Problem Based Learning, *Univ. Chem. Educ.*, **7**(1), 13–24.

Mayer R. E., (2004), Should there be a three-strikes rule against pure discovery learning? *Am. Psychol.*, **59**(1), 14–19.

Meyer H.-D. and Schiller K., (2013), Gauging the role of non-educational effects in large scale assessments, in Meyer H.-D. and Benavot A. (ed.), *PISA, Power and Policy*, Oxford: Symposium Books, pp. 207–224.

Popham W. J., (1999), Why Standardized Tests Don't Measure Educational Quality, *Educ. Leadership*, **56**, 8–16.

Popham W. J., (2010), *Everything School Leaders Need to know about Assessment*, California: Corwin.

Reid N., (1976), Simulations, Games and Case Studies, *Educ. Chem.*, **13**, 82–83.

Reid N., (1978), *Attitude Development Through a Science Curriculum*, PhD thesis, Glasgow: University of Glasgow.

Reid N., (1980), Understanding Chemical Industry - Teaching Materials, *Educ. Chem.*, **17**(3), 78–80.

Reid N., (1982), The Time Dimension in Chemistry, *Educ. Chem.*, **19**(6), 166–168.

Reid N. and Yang M.-J., (2002), The Solving of Problems in Chemistry: The More Open-ended Problems, *Res. Sci. Technol. Educ.*, **20**(1), 83–98.

Renkl A. and Atkinson R. K., (2003), Structuring the transition from example study to problem solving in cognitive skill acquisition: A cognitive load perspective, *Educ. Psychol.*, **38**(1), 15–22.

Robinson K., (2011), *Out of Our Minds - Learning to be Creative*, Chichester: John Wiley and Sons.

Sweller J., Kirschner P. A. and Clark R. E., (2016), Why minimal guidance during instruction does not work: A reply to commentaries, *Educ. Psychol.*, **47**(1), 115–121.

Sweller J., van Merriënboer J. J. G. and Paas F., (1998), Cognitive architecture and instructional design, *Educ. Psychol. Rev.*, **10**, 251–296.

Tsaparlis G., (2009), Learning at the Macro Level: The Role of Practical Work, in Gilbert J. K. and Treagast D. (ed.), *Multiple Representations in Chemical Education*, Springer, pp. 109–136.

Ud-Din M. N., Reid N. and Malik I. A., (2016), *Training Manual on Educational Assessment*, Islamabad: Federal Board of Intermediate and Secondary Education.

van Merriënboer J. J. G. and Kirschner P. A., (2013), *Ten Steps to Complex Learning*, 2nd edn, Hillsdale, NJ: Erlbaum/Taylor and Francis.

Willis J., Willis M. and Huijser H., (2015), Learning Power: Taking Learning-Centredness Seriously in a Blended Learning Environment, in Harmes M. K., Huijser H. and Danaher P. A. (ed.), *Myths in Education, Learning and Teaching, Practices, and Principles*, Basingstoke: Palgrave Macmillan.

Yang M.-Y., (2000), *Problems Solving in Chemistry at Secondary School Science,* PhD thesis, Glasgow: University of Glasgow, http://theses.gla.ac.uk/2161/, Accessed 8 November 2019.

CHAPTER 7

# Tips and Tricks: Practical Ways to Develop Teaching

## 7.1 Looking Back

In the last chapter, we looked at three very broad approaches to teaching:

(1) *Teaching as telling*: the world of didactic presentation, verbally or by means of written or visual material.
(2) *Teaching as sharing:* where learners collaborate and cooperate, often in small groups, to apply chemical ideas to solve a problem or to engage with an issue where chemical insights are important.
(3) *Teaching as doing*: the main element here is the experimental enquiry where learners are exploring the way the molecular world reacts when we ask questions of it.

In all of this, the central goal is to enable learners to make sense of the world. Indeed, the central goal for education has to be understanding. For chemistry, this means understanding how our world works when considered at the molecular level.

We saw that the central problem relates to the fact that the molecular level is inaccessible to humans' senses in any direct sense. In seeking to make sense of that world, in the end we must invoke the sub-micro and the representational. This immediately places the working memory under pressure as it seeks to cope with chemistry at three levels: this is the problem summarised in the Johnstone triangle. Fundamentally, that is why chemistry is so often demanding for learners.

In this chapter, we shall move on to look at some very practical ways that have been found in specific areas of chemistry that can help our learners

Advances in Chemistry Education Series No. 6
The Johnstone Triangle: The Key to Understanding Chemistry
By Norman Reid

Published by the Royal Society of Chemistry, www.rsc.org

forward. The extent of understanding is controlled by the limited and fixed nature of working memory capacity. The extent of understanding is *not* controlled by the teaching method employed. However, specific approaches *within* any method may be helpful in minimising the possibility that the working memory will become overloaded. The key is not the method but how the method is employed.

**Some Key Principles**

The extent of understanding is:

- Controlled by working memory capacity.
- Not controlled by the teaching method employed.
- Controlled by the way the teaching method is employed.

Many of the topics and themes discussed in this chapter arise directly from the work developed by Johnstone in his attempt to minimise the working memory overload problem. This illustrates the approaches he adopted. It is hoped that this will encourage us be much more critical of traditional ways of teaching chemistry at school levels but also be much more creative in seeking new ways forward, based soundly on the evidence that Johnstone demonstrated so clearly and that are summarised in his triangle model.

## 7.2 Atomic Theory

It is a common feature in many school curricula for a course to start with long stretches where the findings of decades of research are brought together to show how atoms are constructed, how electrons behave in terms of shells and pairing, how bonds form, the nature of these various bond types, and the way molecules and ions then arise.

Inevitably, this involves extensive use of symbolisms and representations as sub-microscopic ideas are presented. The Johnstone triangle predicts that working memory overload is almost inevitable. Johnstone argues against this approach and that we should seek to introduce the absolute minimum of such '*atomic theory*'. Indeed, such theory is very often dependent on evidence of isolated atoms from the gas phase, a situation that is well beyond the experience of the young learners. Johnstone *et al.* (1981) present a much sounder approach.

This book was written for school students, aged about 14–15. By this stage, they had already met the words: '*element*', '*compound*', '*atom*' and '*molecule*'. He presented the elements as the building blocks (he used the word '*bricks*') from which the entire physical world is constructed. He then described molecules as '*clusters of atoms*' linked together in some way (Johnstone *et al.*, 1981, pp. 1–2).

He moved on to introduce protons, neutrons and electrons (in the curriculum at that time, the experimental evidence for these was presented in the previous year). Much of this was all revision and was very much centred on the Periodic Table as a useful way to list the elements by the masses of their atoms. It was then easy to refine the idea later to show how the Periodic Table presents the elements in order of atomic number.

At this point, he employed modelling balloons to show the way electrons around atoms will be located. Using these balloons, it was easy to show the locations of electrons between adjacent nuclei. It was easy for the young learners to see how some negative charge caught between two positively charged nuclei could hold the nuclei together. The key stereo-chemical ideas emerge naturally from this model. This entire area is brief and constitutes a very small fraction of the entire textbook. It is linked to the macro wherever possible. The minimum of the sub-micro and representational is introduced.

This entire presentation reflects the key messages from the Johnstone triangle: start with the macro (where the learners can employ their senses) and introduce the minimum of the sub-micro and representational carefully, seeking to avoid working memory overload. The evidence suggests that this approach works really well.

At that point, the book moves into carbon chemistry where the bonding rarely involves the ionic and where vast amounts of important chemistry can be unpacked without employing any more from the world of the sub-micro or representational. There is one great additional advantage: the students were able to practise the use of simple symbolisms, therefore gaining confidence in these. This includes simple formulae and equations and the way these can be used to summarise what the students can observe in the experiments they undertake. Much simple organic chemistry can be handled meaningfully in this way.

The overall sequence of ideas in the book is shown in Figure 7.1.

Only after all this does the book proceed to look more at the behaviour of electrons in bonds. This leads to the ionic situation. This moves the book into the inorganic areas, much being built around the chemistry of sulfur and nitrogen. Only at the very end of the two-year course were acids and alkalis formally introduced and the dreaded mole considered.

The book reflects the chemistry curriculum at that time in Scotland although many countries followed similar topics. What Johnstone had done was to re-order the way it was presented. The limitations of working memory capacity were foremost in his mind (Johnstone *et al.*, 1981, p. ix).

At the same time, he did not introduce misleading (and often totally wrong) models and ideas like planetary electron movements, octets, atoms '*wanting*' to give or receive electrons. He never made the assumption that a bond has to be two electrons or that covalent and ionic binds are two different things (in fact varying degrees of polarity is perhaps better to be seen as the norm). Only the existence of lattice energy or solvation energy makes

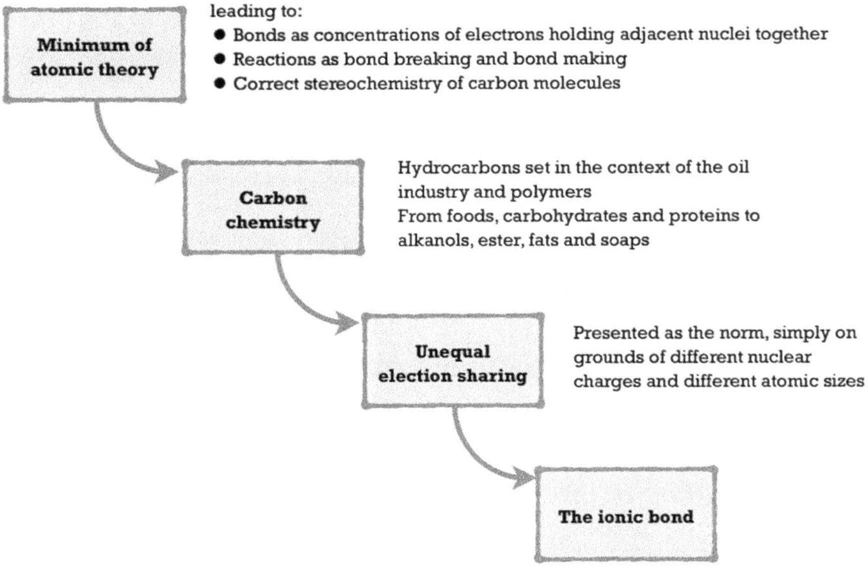


**Figure 7.1** Teaching sequence based on the Johnstone triangle.

an ionic bond so common. Gaseous sodium chloride (hardly a common experience!) has been shown as perhaps best represented as an ion-pair or, in other words, an exceptionally polar bond.

## 7.3 The Mole

Early in his research, Johnstone found that the mole concept posed immense problems to school learners and even with undergraduates (Johnstone *et al.*, 1971; Duncan and Johnstone, 1973; Johnstone, 1974). Although his work related to students being taught in Scotland, the problems seem to be worldwide. Arising from this early work, it became apparent that:

(a) The mole was linked in the minds of '*novice learners*' to mass.
(b) The mole linked to gram formula mass was accessible to most learners.
(c) As soon as the mole was linked to solutions, the problems grew rapidly.

In the work of Duncan, it was shown that students about age 15–16 could link mass to numbers of moles. However, even apparently straightforward calculations like: '*How many* $cm^3$ *of 1.5 M NaOH would be required to neutralise 25* $cm^3$ *of 2.6 M* $H_2SO_4$*?*' caused problems.

We saw how the mole in aqueous solution was left right to the end of the two-year course (ages 14–16) as presented in his book (Johnstone *et al.*, 1981, pp. 205–209). Earlier (Johnstone *et al.*, 1981, pp. 124–126), he had introduced

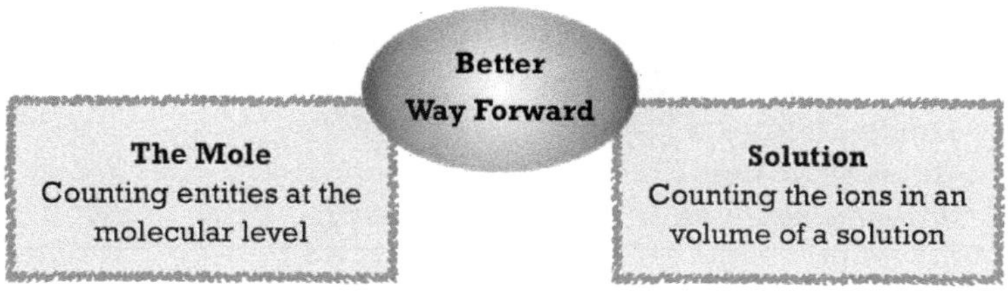


**Figure 7.2** Some improvement.

the mole as a *counting mechanism* to relate the molecular world to the macro world of real quantities of materials. In passing, this is typical of the strength of the '*spiral curriculum*' advocated by Bruner decades before (Research briefs, undated). Here an idea is touched upon briefly to be developed further at a later stage.

Let us stop and reflect on the mole as a counting mechanism. It is very difficult for the novice learner to conceptualise the incredibly small size of entities at the molecular level. For example, in a teaspoonful of water, there are about $1.67 \times 10^{23}$ molecules of water. Even writing this out in full will leave most young school students with glazed eyeballs.

An interesting task can be given to a class. Working in small groups (of, say, three students), set the group the challenge to estimate the number of moles of sand grains in the Sahara Desert, given its area as 9.2 million $km^2$. They have to suggest a possible average sand depth and the likely volume of a sand grain. The outcome certainly shows the sheer size of the mole as a number. Here is one way to make the sheer size of the number that is the mole more tangible. It is also (indirectly) a way to make the learners think about the incredibly small size of entities at the molecular level.

The Johnstone approach, when introducing volumetric ideas, was to build on this counting idea and encourage the learner to look at any solution and use the mole as a way of counting the ions present in a volume of solution. Simple acidimetry calculations then followed. This approach seemed to help a bit but there were still problems (Figure 7.2).

A similar approach was suggested in a teachers' manual published slightly later (Reid, 1982). The manual was published as a guide to teachers following a revision of the course that followed the Higher Grade Chemistry course in Scotland.† The manual offered three ways to approach volumetric calculations. The first two reflected the earlier Johnstone approach but extended to apply to redox as well as compleximetric calculations. The third approach was rather different and had an interesting history.

†By an accident of history, the university qualifying examinations are taken in the penultimate year of secondary education (although more Higher Grades can be added the following year). This leaves the aim of the final year as an overt preparation for university life. Certificated course all include individual projects to this end, allowing more independent work.

**Scottish Examination Structure (around 1980)**
Ordinary Grade Chemistry (2 years)
Higher Grade Chemistry (1 year)
Certificate of Sixth Year Chemistry (1 year)
*[University entry based on Higher Grade passes.]*

During a train journey to a chemistry meeting, Johnstone was involved in a discussion about the apparently intractable mole problem. It was the time of year when it was taught at three different levels in Scottish schools:

| | |
|---|---|
| Age 15+ | Simple acidimetry towards the end of the course covered in Johnstone's book. |
| Age 16+ | More complex acidimetry and simple redox ideas. |
| Age 17+ | Redox, gravimetric and compleximetric experimental work and calculations. |

Arising from that conversation came the idea of '*power*'. Power could be defined straightforwardly (Table 7.1).

This gives a simple unifying concept. However, it allows all volumetric calculations to be '*chunked*' in terms of an easily memorisable formula. This has the scope to reduce the pressure on limited memory capacity and enable learners to gain confidence in volumetric calculations: it was found that using this approach, the students tended to obtain correct answers with an amazing frequency! They managed this with relative ease and confidence was observed to grow.

We must never underestimate the power of perceived success to generate confidence. In a study in a very different cultural setting, Oraif (2007) found that the only factor she could detect that related to academic confidence (at first year university level, reflecting school experiences) was perceived success in the past. The evidence shows that using the idea of '*power*' generated confidence and the was based on success. The unifying concept of '*power*' approach was described briefly in Reid (1982) and was later re-published in Reid (2010).

**Table 7.1** The concept of '*Power*'.

| Defining 'Power' | |
|---|---|
| Acidimetry | The number of hydrogen ions produced or absorbed by one molecule of the acid or base. |
| Redox | The number of electrons produced or absorbed by one molecule or ion. |
| Compleximetry | Defined in terms of coordination numbers and denticity. |

If we think about the ability of an acid to neutralise a base, three factors must be involved:

(a) The volume of acid used.
(b) The molarity of the solution.
(c) The '*power*' of the acid to neutralise.

If *power* is defined as the number of hydrogen ions produced or absorbed by one molecule of the acid or base, then both hydrochloric acid and sodium hydroxide have a power of one while sulfuric acid and calcium hydroxide have a power of two. This requires understanding of the meaning of the formulae.

This leads to a simple relationship:

$$\underset{(acid)}{V_1 \times M_1 \times P_1} = \underset{(base)}{V_2 \times M_2 \times P_2}$$

This brings together volume ($V$), molarity ($M$) and power ($P$). It is found that using this relationship enables students to gain right answers. The question is whether this is being achieved at the expense of understanding. In fact, this is not so: the reaction has to be understood *before* using the concept of '*power*', as we shall see.

Consider the following problem:

*Calculate the molarity of a sodium hydroxide solution if 25 mL is exactly neutralised by 10 mL of 0.1 M sulfuric acid.*

The student needs to know the formula of the acid and base and, hence, deduce the '*power*' of the acid as two and the '*power*' of the alkali as one. Balancing the equation is important. The method has the great advantages in that it gathers all the variables into one easily remembered relationship which can be applied in a straightforward fashion. This is an example of chunking (Miller, 1956).

**Understanding is Critical**
If an acid like ethanoic acid (acetic) is used, the student has to *understand* the reaction and that only one hydrogen is involved in the formation of water.

When this approach was applied with students, it was found that they coped well on meeting neutralisation for the first time. They almost invariably obtained right answers and did not view the topic with dread. When the approach was adopted with older students who been taught previously by different ways, there was an interesting reaction, typified when one student said, '*Why did you not teach this the way last year? It's easy.*'.

However, there is another even greater advantage. The same relationship applies to redox reactions as well. Here, '*power*' is seen in terms of the *electrons lost or gained per molecule or ion of reactant.*

Consider the following analysis (cited verbatim from Reid, 2010, pp. 43–44).

*If 20 mL of 0.02 M potassium permanganate is acidified and treated with excess potassium iodide solution, iodine is released. When this iodine is titrated against 0.1 M sodium thiosulfate solution, using starch indicator, what volume of thiosulfate would be used?*

Ion electron equations have to be developed and balanced:

$$MnO_4^- + 8H^+ + 5e^- \rightarrow Mn^{2+} + 4H_2O$$
$$2I^- \rightarrow I_2 + 2e^-$$
$$2S_2O_3^{2-} \rightarrow S_4O_6^{2-} + 2e^-$$

| |
|---|
| This ***demands*** that the students understand the relevant chemistry. |

From these, the relevant values for '*power*' are easily observed under these conditions:

Power of permanganate ion = 5
Power of iodine molecule = 2
Power of thiosulfate ion = 1 (note: electrons per molecule or ion)

The whole calculation can be done easily:

$$\underset{(permanganate)}{V \times M \times P} = \underset{(iodine)}{V \times M \times P} = \underset{(thiosulfate)}{V \times M \times P}$$

The student can see easily that all that is need is the permanganate–thiosulfate relationship, the iodine not being relevant. The calculation reduces to:

$$\underset{(permanganate)}{V \times M \times P} = \underset{(thiosulfate)}{V \times M \times P}$$
$$20 \times 0.02 \times 5 = V \times 0.1 \times 1$$

giving the volume of thiosulfate as 20 mL.

It was found that students aged about 17+ found this approach easy. At that time, the Scottish Certificate of Sixth Year Studies chemistry paper offered eight questions at 20 marks each and the candidates had to select five. After having been taught this approach, the question based on volumetric calculations was chosen by many students, whereas, as a rule, very few students nationally were observed to opt for the volumetric question.

Look at the follow typical examination question:

> *25 mL samples of a solution containing $Cu^{2+}$ were treated with excess potassium iodide and the iodine released titrated with 0.1 M sodium thiosulfate solution using starch indicator, 8.25 mL being required. Calculate the molarity of the original copper solution.*

Any analysis of this question shows far more ideas being required to be held *at the same time* to gain success. Working memory overload is virtually guaranteed. However, the approach using the concept of '*power*' is found to work well. However, we do need to teach the students to follow a simple procedure:

> Students need to ***understand*** the chemistry involved.

(1) Write down the relevant ion–electron equations and balance them (we shall show later a simple way to approach this).
(2) Look at the electrons produce or generated per molecule or ion and this deduce the relevant '*powers*'.
(3) Apply the formula.

In fact, many students, looking at the equations, quickly realise that the iodine released is a '*go-between*' in that the Cu(II) ions do not react directly with the thiosulfate ions. The electron transfers are what is important. This is sound chemistry: the electrons hold the key in any chemical reaction.

Thus:

$$\underset{Cu^{2+}}{V \times M \times P} = \underset{I_2\,/\,I^-}{V \times M \times P} = \underset{S_2O_3^{2-}}{V \times M \times P}$$

Is reduced to:

$$\underset{Cu^{2+}}{V \times M \times P} = \underset{S_2O_3^{2-}}{V \times M \times P}$$

## 7.4 Redox

In most countries, the area relating to redox comes in the later years of school chemistry studies. Many students find this totally confusing as Johnstone *et al.* (1971) demonstrated. In one paper, Garforth *et al.* (1976a) found that ion equations were simply not understood by younger learners but things did improve by about age 17 +. In the second paper, Garforth *et al.* (1976b) analysed the underpinning concepts and suggested an order of presentation that helped. It has to be recognised that these studies pre-dated the triangle model, but the critical importance of working memory was now beginning to be appreciated. Garforth had noted that, to reach a simplified summary ion–electron equation, learners often had to go through complex, high information stages. Success was elusive.

There is a further problem. Too often, redox equations are seen to involve some kind of artificial mental gymnastics. The symbolisms are left unrelated to what the learner can observe in a test-tube. We shall return to this issue later.

For the moment, let us return to the actual ion–electron equations. What the learners need is some simple strategy to handle these. Once that strategy is established, the meaning can be drawn in and the representations linked more tightly to the macro chemistry involved.

A simple way was developed to help students balance ion-electron equations. This involved four steps:

(1) Balance atoms other than hydrogen and oxygen.
(2) Balance oxygen by adding water as necessary.
(3) Balance hydrogen by adding hydrogen ions ($H^+$) as necessary.
(4) Balance charge by adding electrons ($e^-$) as necessary.

This series of steps is easy to remember. Elements, oxygen, hydrogen, charge. It is another classic example of finding a way to avoid working memory overload. There are far too many variables for the working memory to handle *at the same time* when balancing redox equations. By teaching the learners to focus in on one variable at a time, the balancing becomes quite straightforward. However, what the young learner needs at the outset is the confidence derived for success. Here, the order of balancing is important.

For final year school chemistry students, Johnstone developed the iodine coulometer as a way to link the electron transfer in an ion–electron half equation to the physical reality of the experiment.

No constant source of direct current is needed nor is any sophisticated cell. Figure 7.3 illustrates the apparatus that is needed (Reid, 1982, p. 4).

2 mL of 0.1 sodium thiosulfate solution is added to the beaker and the current passed until the iodine colour just persists (starch). The value of the current (need only be approximate) and the time required are both noted.

Number of electrons passed = Current × time/charge on each electron [Amps × seconds/coulombs per electron]

$$= I.t / 1.6 \times 10^{-19}$$

$$= (I.t) \times 6.25 \times 10^{18}$$

Number of thiosulfate ions used $= (2 \times 10^{-3}) \times [(0.1) \times (6 \times 10^{23})]$

$$= 1.2 \times 10^{23}$$

The values for current and time are inserted into the first equation. A comparison of the two numbers shows that one electron is involved in the reaction of a thiosulfate ion. The actual measurement does not require any precise accuracy for the answer can only be a whole

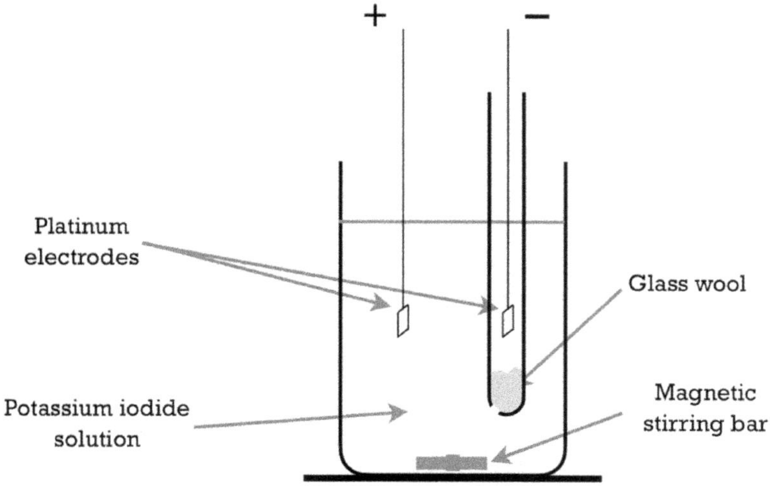


**Figure 7.3** The iodine coulometer.

number. There are numerous steps involved in this calculation and the teacher may need to offer assistance to the learners. Nonetheless, in this way, some macro evidence is related to the sub-micro and representational.

A fresh way of looking at ion equations is to look at the chemistry of manganese. Three manganese compounds are usually available in most schools. Bottles of two of these can be placed before senior school students: manganese(II) sulfate and potassium permanganate. These feature the lowest and highest oxidation states. The students can now be challenged to search in books or the internet to find all the oxidation states of manganese and then to find ways to generate a sample of each starting from one of the two compounds.

They can collaborate in this, dividing the work up, before demonstrating to each other the appropriate reactions: all can be done in test-tubes. Placing the six oxidation states side by side reveals a range of colours. Writing the ion–electron equations for the reactions links oxidation state to electrons transferred very neatly. It also illustrates the importance of hydrogen ion concentration in the various reductions starting with the permanganate, justifying pH control. Here is another simple way to link the macro to the other two levels. It has two other bonuses: it presents students with simple experimental challenges and, at the end, offers great scope for discussion as many aspects of transition metal chemistry are explored.

## 7.5 Thermodynamics

Here is an area where it is very easy to present the topic in terms of abstract mathematical representations. Very often, the macro chemistry can almost disappear as students are taught how to select and then apply mathematical manipulations. Indeed, in some presentations the chemistry almost seems to disappear as the area of study develops a mathematical life of its own!

At the outset, Johnstone linked the ideas of exothermic and endothermic reactions tightly to the formation of chemical bonds. Therefore, an overt link was formed between the macro (observed energy changes) and the representational (bond breaking and formation). Much was built on the simple ideas that bonds have to be broken before new bonds can be made and that tied back into the ideas underpinning kinetics and catalysis. This simple idea of bond-breaking followed by bond-making underpinned the approach adopted at earlier stages and was seen in Johnstone's textbook for younger ages (Johnstone *et al.*, 1981). Again, this illustrates the importance of being very careful at earlier stages to lay the correct foundations to enable future learning to build on easily.

In all this, it is important to link things back to the macro. In one book (Johnstone and Morrison, 1969, p. 27), there is a description of a simple experiment that illustrates how the calculations side of chemistry can be linked back tightly to the macro and then symbolic. Some simple apparatus was constructed (Figure 7.4).

Screw top jars are used as burners and a pipe cleaner acts as a wick in each. A different alkanol (alcohol) is placed in each: methanol, ethanol, propan-1-ol, butan-1-ol. The burners are weighed before lighting. The heat energy released is measured by looking at the temperature rises in a series of beakers, each containing 100 mL (or smaller volume) water, the temperature being measured using a stirring thermometer.

The burners are lit, and the temperature rises in the water observed, stirring periodically. When the temperature in a beaker rises by 10 °C, that burner is extinguished. The burner is then re-weighed to find out how much alcohol has been used to generate the heat to raise the temperature of water by 10 °C. Of course, there are all kinds of heat losses but all we are interested in is a *comparison*, for the heat losses will be similar for all four beakers.

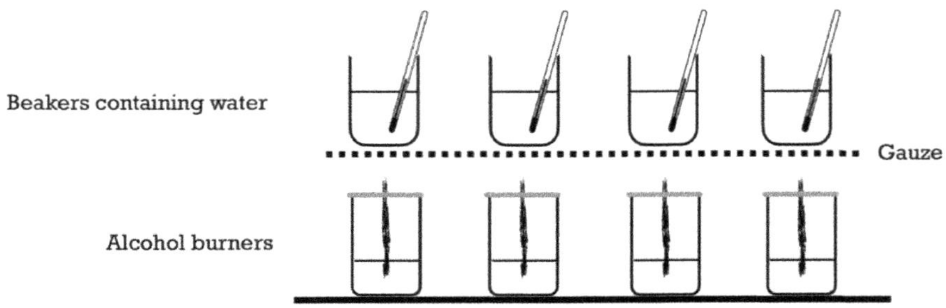


**Figure 7.4** Heats of combustion.

The heat transferred to the water will be 4.2 kJ (100 mL at 10 °C rise). The masses of alcohols burned can be found. It is, therefore, easy to calculate the heat energy related by burning a mole of each alcohol. While the figures will not be the same as the figures in data books, the learners can see the trend in heat energies released and link that back to the structures and the combustion reaction. In fact, they will see an almost constant difference from one alcohol to the next.

Here is a simple way to link the macro to the sub-micro (the structure and bond-breaking and bond-making stages), thus making this area of thermochemistry a little more real for the learners. However, Johnstone went much further in his attempt to enable learners to link the macro with the representational (mainly mathematical) in more formal thermodynamics (Johnstone and Webb, 1977). Everything in the book links the ideas of thermodynamics back to macro chemistry. Chemical reactions are very often exothermic. This is linked to the breaking and making of chemical bonds. The bonds formed release more energy than the energy absorbed when bonds are broken. Overall, the system releases energy to its surroundings. This makes immediate sense to learners: the system is losing energy.

The problem arises when we look at reactions which are endothermic and yet proceed quite happily on their own. Here are some examples where we can allow the students to undertake the experiments and see if they can spot any common features: reaction of thionyl chloride with hydrated cobalt(II) chloride in a fume cupboard, shaking hydrated barium hydroxide crystals and ammonium thiocyanate crystals together, reaction of ammonium carbonate with propanoic acid. Each of these reactions proceeds quite happily on its own, absorbing heat energy from the surroundings, the temperature falling steadily.

The question posed to the students is: why a reaction should be '*willing*' to take place when it takes in energy: it is like seeing a car moving uphill on its own. The teaching can then lead them forward to see endothermic reactions that proceed on their own as situations where greater disorder is created. Could this be an explanation?

## 7.6 Entropy

The concept of entropy is introduced and a simple way to combine enthalpy and entropy is generated. Now we have free energy, and this is the key factor determining whether a reaction will proceed or not on its own. The first and second laws of thermodynamics just arise naturally and the learners quickly appreciate the importance of the second law in determining the direction of change. The concept of reaction feasibility is thus established. At that stage, free energy can be linked to the established ideas of equilibrium and the added factor of activation energy is integrated.

The central equation throughout is:

$$\Delta G^0 = \Delta H^0 - T\Delta S^0$$

While the ideas have developed with the macro chemistry always held in view, the book (Johnstone and Webb, 1977) finishes by applying the ideas to a whole range of practical situations: diffusion, osmosis, concentration cells, separation process, ion exchange, solution processes as well as metal extraction and photosynthesis.

Johnstone had a neat way to bring ideas together. He placed a petri dish containing super-saturated sodium ethanoate on an overhead projector platform. He dropped in a single crystal of sodium ethanoate and the students could then see the rapid formation of crystals, spreading out. Here is case where there appears to be increased order taking place on its own. He then took the petri dish off and placed on the upturned palm of a nearby student. The student immediately would say, '*It's hot!*', and pull their hand away. This opened the door to the important emphasis in considering the direction of any spontaneous change: *overall* entropy must increase and this must take into account the system *and* the surroundings. In these simple ways, Johnstone employed the experiment to link the macro to the mathematical representations. These approaches are developed in Johnstone and Webb (1977, pp. 22–48).

Further ideas were developed, and some can be seen in Reid (1982, pp. 14–15). For example, the concept of entropy is introduced here using a milk bottle analogy. At that time, milk was sold in bottles and the cream tended to rise to the top. The teacher would take the bottle and upturn it once and place it back on the front bench. The students could physically see that some mixing had taken place. They can deduce that this had been caused by the teacher expending some energy in upturning the bottle once.

One physical upturn creates some disorder. The more upturns the more disorder is created. Thus, increase in disorder (cream and milk mixing) is proportional to energy input (the upturns).

In formal terms:

$$\Delta S \propto \text{energy supplied}$$

or

$$\Delta S \propto q$$

However, the amount of disorder depends on whether it is the first upturning or a later upturning. The students are aware that temperature rises generate more movement at the molecular level and this will create more disorder.

In formal terms:

$$\Delta S \propto 1/T$$

We can bring this together:

$$\Delta S \propto q/T$$

or

$$\Delta S = q/T$$

(if units are chosen correctly)

This gives the units for entropy change as energy units per degree: Joules per Kelvin.

If we wish to relate this to a specific amount of matter, then we have the final relationship:

$$\Delta S = \Delta H/T$$
$$(\mathrm{J\,K^{-1}\,mol^{-1}})$$

It is then easy to combine two energy terms to generate the equation:

$$\Delta G^0 = \Delta H^0 - T\Delta S^0$$

All these approaches encourage the mental linking of the macro with the mathematical representations of what is going on. This is another example of Johnstone's triangle in operation where the macro (this time the then very familiar cream-milk situation) was employed to make the representational meaningful.

The ideas of entropy can be illustrated further. The students need to grasp the second law: in any spontaneous process, overall entropy must increase.

A spontaneous process is defined as any process that can take place on its own, irrespective of how long this may require.

At this point, it is useful to allow students to look at the entropy values for a few gases, liquids and solids. There is no need to discuss how the values were measured. It merely gives the students a '*feel*' for the differences between gases, liquids and solids.

This can lead to the development of a graph (Figure 7.5) to illustrate entropy increases for any substance (just seen as broad trends) (Reid, 1982, p. 13).

We can illustrate the probability basis of entropy by placing four white balls and four red balls in a bag. The chances of pulling the balls out in any

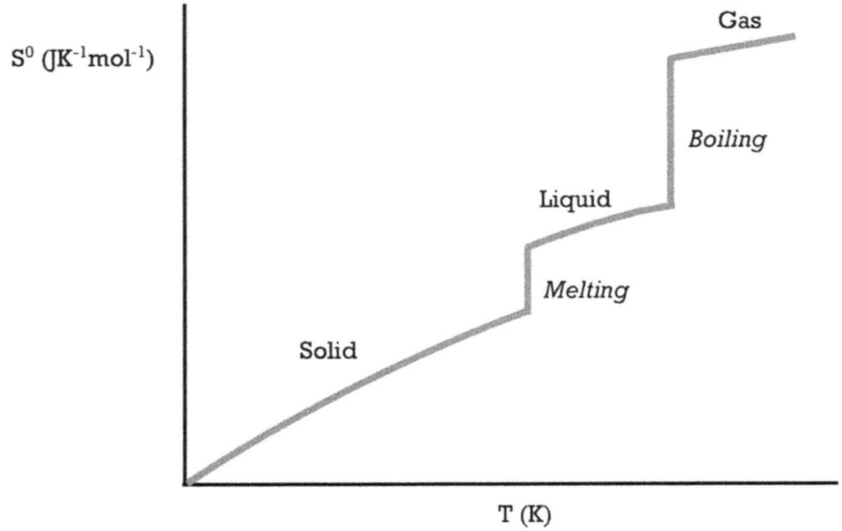


**Figure 7.5** Entropy and the physical states of matter.

ordered fashion (*e.g.* four red followed by four white) are low. This follows simply because there are far more disordered arrangements. If we translate that to the scale of a mole of molecules, it is easily apparent that the chances of any ordered arrangement are so unlikely as to be near impossible. Disorder at the molecular level has to be the norm.

This can then be illustrated by referring to the untended garden, the corrosion of metals, the dispersion of sewage in the sea, the state of a teenager's bedroom (without the effect of a nagging mother), the span of life, *etc.* We can go further to show how we tell the time by the inexorable move to disorder. If a film is shown backwards, then we know it is backwards because there is a spontaneous increase in order—an impossibility. In fact, if there is no visible change in disorder, then we cannot tell if a film is running forwards or backwards: the video of a child on a swing (provided it does not run too long), shows no visible entropy changes. In all these ways, the concept of entropy can be made real at a macro level.

## 7.7 Ideas for Younger Learners

In his earlier books, Johnstone developed numerous simple ways to help younger learners to make sense of chemistry.

Very early in their studies, he would add salt to water, something that most young people have seen happening many times in their kitchen at home. However, Johnstone used a measuring cylinder. Shaking caused the salt to dissolve and the students were asked to look at the water level before and after. Where had the salt gone?

He then took a litre beaker, containing many small balls. Sand was added. Does the level rise? Figure 7.6 illustrates his approach.

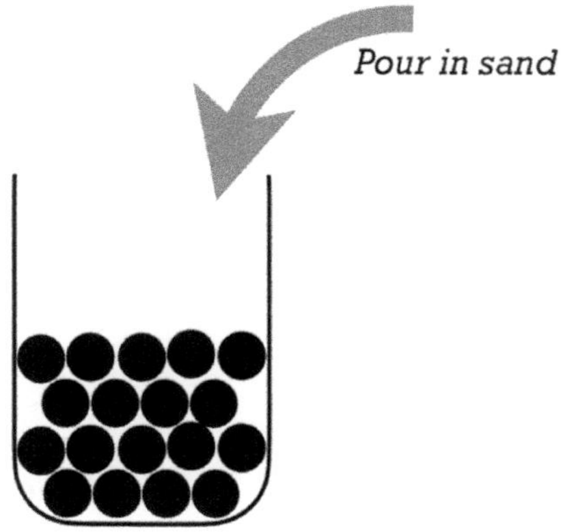


**Figure 7.6** Dissolving salt in water.

In this simple way, he used the macro to develop a mental picture of what might be happening at the sub-micro (Johnstone and Morrison, 1964, p. 4).

He would continue on by setting up a microscope with a slide with a drop of water and a small amount of strongly coloured (*e.g.* blue) poster paint. The learners were asked to look down and describe what they saw. Simple Brownian motion was seen (the name need not be mentioned). Class discussion followed as the students tried to explain the random movements they observed.

Many curricula spend time encouraging students to distinguish between physical and chemical changes. This is an utterly pointless exercise. With overcrowded curricula, here is one area which can be ejected easily. Let us return to our salt dissolving in water. Is this a physical or a chemical change? It is relatively easy to reverse. There is no detectable energy change. However, bonds have been broken and new bonds formed but the energy balance gives little discernible energy change overall.

By placing the emphasis on the supposed different criteria for physical and chemical changes, we make it harder for ideas like equilibrium (especially solution equilibria) to make sense later. This illustrates the key principle that, in our attempts to make sense of the world around, we need to be careful not to develop mental models and understandings that need to be un-taught at later stages.

In considering the physical world around, it is possible for very young students to grow crystals. To gain good outcomes, seed crystals need to be tied on to a nylon thread and suspended in saturated solutions in containers in a thermostatic water bath. It is not too difficult to grow crystals of copper(II) sulfate, nickel(II) sulfate and the alums where the crystals are 2–3 cm long. That is the world of the macro. It relates to the world of crystals and gemstones.

However, the question is why such materials form such beautiful crystals and why certain shapes keep re-appearing. Johnstone (1964, p. 12) used simple models to illustrate the formation of patterns (Figure 7.7).

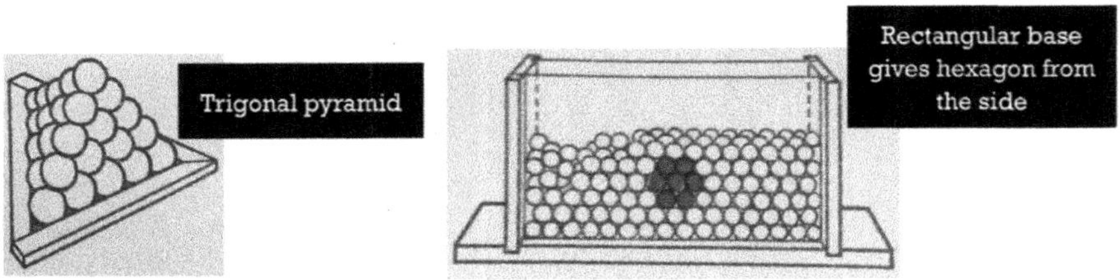


**Figure 7.7** Formation of lattices.

The raises all kinds of discussion points. Are entities at the molecular level to be seen as spheres? What happens if they are not? Are they physically touching as the model suggests? The sand dropped into a beaker containing small spheres suggests space between molecular entities. What happens if there are two different kinds of entities, of different size? What happens then? The macro has been brought together with the sub-micro in a meaningful way.

In his early career, Johnstone would describe himself as an analytical chemist. He had a fascination for qualitative inorganic analysis. This can be seen as rather like the world of the detective and, indeed, forensic science depends heavily on very sophisticated analytical methods. He introduced the flame test early as a way to detect a few elements. This can be derived from the observation of a yellow-orange colour given to a flame when the saltwater in boiling potatoes boils over.

The students can undertake these with unknown samples, using silica rods instead of platinum wires. Here is careful observation at work. Students develop ways to ways to mask the dominant sodium yellow-orange to see some of the more delicate colours. They start to see that some metal ions (*e.g.* calcium) generates its fleeting brick-red colour *later* than the sodium yellow-orange. There are numerous practical applications including older street lighting, spilling salt on a gas flame, to fireworks. Many decades ago, a course called '*The Chemical Detective*' was developed for less academic students aged about 14–15. This showed just what could be achieved in the area of macro-chemistry (pre-dating the fashion for TV shows based on the forensic). Here is rigorous macro chemistry at work.

In Chapter 5, we saw the way trends in specific properties can be related to the Periodic Table, using graphs of atomic number against the property. It is possible to look at the occurrence of the top ten elements in terms of a pie chart. This shows very vividly how ten elements dominate the make-up of the accessible part of our planet. A variety of questions can be discussed by young students, working in a small group. Alternatively, the teacher can invite questions from the class (Figure 7.8).

Here is macro chemistry at work. All kinds of questions can be generated by this simple pie chart. It is even possible to mount a coloured pie chart as above on the wall of the teaching area. Sooner or later, the students start to ask questions that can lead on to an informal but valuable discussion.

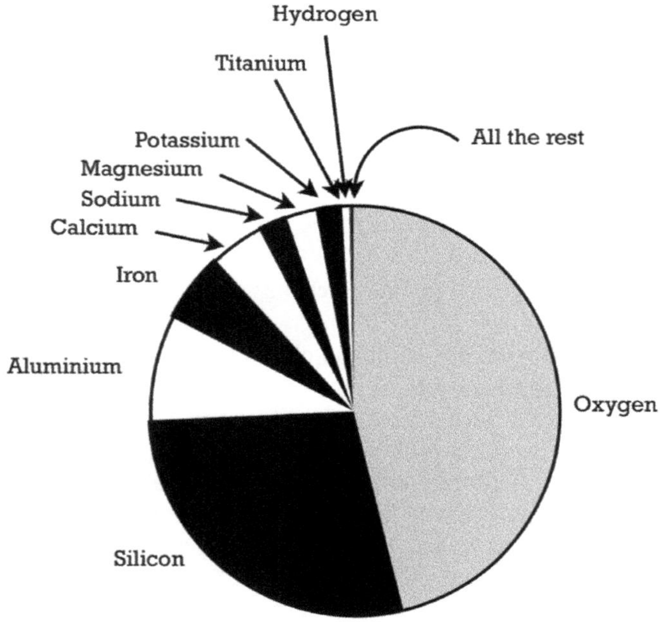


**Figure 7.8** Elemental occurrence.

## 7.8 Organic Chemistry

In his early work, Johnstone had found that some areas of organic chemistry posed difficulties for learners. Hydrocarbons posed few problems, but the difficulties started to arise with reactions like esterification and with the chemistry of alkanals (aldehydes) and alkanones (ketones). One of the key problems is that students often do not see organic molecules in terms of chunks: in other words, functionality is not yet established. Part of the difficulty lies in the way organic chemistry is normally introduced, with hydrocarbons first. Much of the reactivity of organic compounds rests in the functional groups, not the carbon skeleton. Yet we start by focussing the skeletons (Figure 7.9).

There is good argument for starting organic chemistry by looking at, say, the alkanols. This would establish that reactions relate mainly to the hydroxyl group. This could then move forward to primary, secondary and tertiary alkanols and then on to alkanals and alkanones. However, this is not the normal way teaching sequences are presented.

It is worth considering starting with alkanols rather than hydrocarbons in that this lays much greater stress on functionality and the way functionality is central in understanding reactions with organic compounds. However, this may generate high working memory demands and care will be needed. Here is an area where some careful research is needed.

The central problem, as always, relates to working memory. Learners need time and experience to be able to see organic molecules in terms of carbon skeletons, with correct stereochemistry. Learners need time and experience

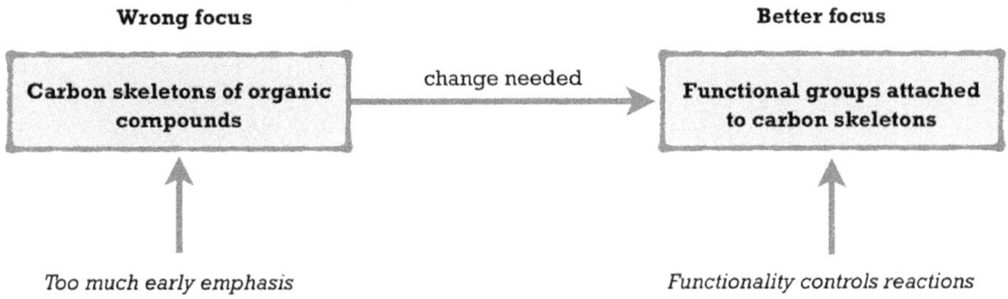


**Figure 7.9** A better approach to organic chemistry.

to be able to see functional groups as clear entities. If the time and experience is not provided, then too much working memory capacity is taken up trying to envisage the structures. Achieving any *understanding* of what is happening in the reactions is thus highly unlikely.

However, there is a further working memory problem. On paper, many organic molecules are pressed flat, distorting the stereochemistry. Representation and reality are not consistent adding another burden to the working memory. Molecular models, coloured pictures and the various ways by which three-dimensional structures are represented on paper need to be provided and time allow for the learners to bring them together so that structures can be '*seen*' mentally.

Overall, it is not easy to make sense of organic reactions and why they occur until the key ideas of structure, bond polarity and stereochemistry are well established and are secure in the learners' minds. This frees working memory space to think about the mechanisms of organic reactions and this holds the key to understanding *why* they behave in the way they do.

Hassan *et al.* (2004) explored this. With a large first year university chemistry class ($N = 367$), the understanding of the key foundational ideas was measured, using structural communication grids. This format of testing is powerful in measuring conceptual understanding (Egan, 1972; Johnstone *et al.*, 2000). The measurements were then related to the student performance in their first university course in organic chemistry. This course followed a mechanistic approach. Four key foundational ideas were explored: the nature of the chemical bond, polarity, functionality and stereochemistry.

It was found that bond polarity was the area of greatest difficulty, with problems also arising from the student understandings of functionality and stereochemistry. One of the problems is the way bond polarity is often *added* on after the ideas of covalent and ionic bonding are developed. The Johnstone approach at the outset was to conceptualise a bond as negative charge caught between two adjacent positive nuclei, thus holding the nuclei together. Polarity was then much easier to be understood. By starting with covalent and ionic binding presented as two *very different* ways, polarity is less accessible, but this approach was (and probably still is) the normal way. Induced polarity, so important in so many organic reactions, is one step even more confusing.

However, it was also found that functionality was a problem. Placing the emphasis on the carbon skeleton at the start of understanding organic chemistry did not help. Functionality is a great chunking device, but the early presentation must develop this and lay emphasis on it, allowing time for school learners to gain confidence. Functionality must be *automated* in the minds of learners if it is to act an effective chunking device.

Johnstone and Kellett (1980, p. 178) had demonstrated the importance of chunking skills in understanding organic chemistry. They found that only 14% of senior school students were capable of using simple functional groups (like $CH_3$, $CH_2$, OH, C=O, COOH, $NH_2$) as chunking devices. This made making sense of an esterification equation almost impossible for the majority of the students.

Structural ideas were also found to be very important. The way the university course tends to represent organic molecules on paper using approaches that allow the stereochemistry to be much more apparent. Perhaps, this is an area where the approaches adopted in earlier school course needs further examination to see if there are better ways to enable learners to visualise the real shapes of organic molecules. Perhaps, the is one place where school courses need to look again at university approaches. It is important that we avoid the simplifications of organic structures that can lead to wrong mental ideas. These will cause later problems. The stereochemistries of organic molecules can often be critical in making any sense of reactivity.

The key thing is that the fundamental ideas of bond polarity, functionality and stereochemistry need to be established before a mechanistic approach is adopted. Once established, the working memories of students can the cope with the representations integral to any mechanistic approach. Organic chemistry can then become a highly elegant area of study where students derive great satisfaction from seeing fundamental principles in operation. They are able to not only understand what is happening, linking the macro level to the sub-micro level, with the representation systems secure and meaningful.

## 7.9 Key Messages

The central focus of this chapter has been on some practical ways by which the macro level has been linked to the sub-micro and the representational. The goal here is to illustrate approaches by which the restrictions related to the limited capacity of working memory can be minimised. Learners do not have sufficient capacity to handle all three levels *at the same time.* Johnstone presented all kinds of ways to assist the learners and many other ideas have been developed from them.

It is important to remember that education in any subject area, including chemistry, must be centred on the development of understanding. School chemistry education should never be designed to fill the heads of learners with chemical information. Nor should school chemistry education ever be considered as a subject where we only consider the education of the chemists of the future. Education in and through chemistry is important for every member of any modern society. We are seeking to develop a population that understands something of the great contribution chemistry has made in the development of societies. This has to focus on generating an understanding about how the molecular world works.

In this, the role of working memory is central. This is the part of the brain where understandings are developed. Chemistry education must work in line with the way the working memory operates, a point made so eloquently by Johnstone (2000). Johnstone made major contributions in the four areas identified in Figure 7.10.

In one of his books, Johnstone *et al.* (1981) altered the order of presentation quite radically. He also removed all kinds of misleading over-simplifications that were known to generate later confusions in seeking understanding. All of this was done deliberately to reflect the research evidence available at the time relating to the central role of limited working memory capacity. He presented this overtly in the preface, written for teachers.

He modified the entire approach to the development of atomic theory. First, he reduced the amount that was introduced at the outset. Second, he removed wrong ideas like electrons orbiting nuclei in atoms octets and electrons '*wanting*' to form octets or to leave an atom. He removed the most unfortunate polarisation between covalent and ionic binding and the early insistence on electrons supposedly '*wanting*' to pair up to form electron pair bonds.

The bond was simply conceptualised as a concentration of negative charge holding two adjacent nuclei together. This proved to be a much sounder basis. It had the simplicity and flexibility that allowed more complex insights to be built on to it easily at later stages. He introduced the mole as a counting mechanism: it is a bridge between the real world at the molecular level and the world of representation by means of formulae and equations. He left the area of the volumetric until as late as possible and developed new approaches to help students to make sense of volumetric calculations. Later developments have expanded this considerably.

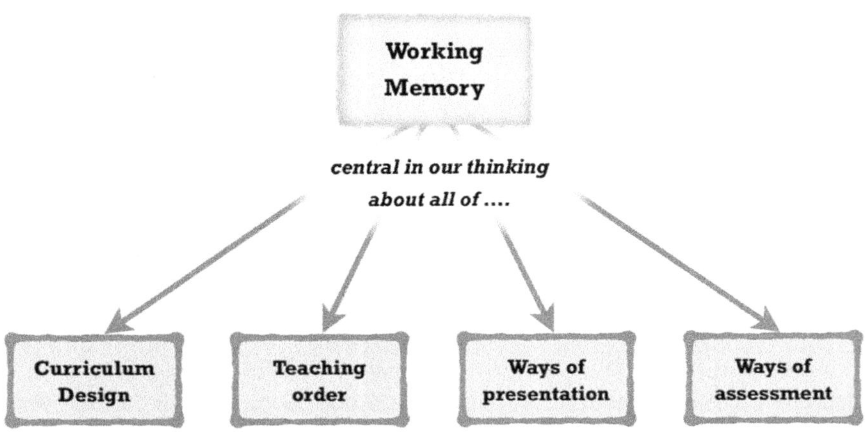


**Figure 7.10** The central role of working memory.

The concept of '*power*' has been found to be a powerful chunking device and, later, can be applied easily and successfully to redox situations. Simple ways were presented to enable students to write and balance ion–electron situations, and to relate what they were writing to the macro world of the experimental observation.

In his work in thermodynamics, Johnstone and Webb (1977) have offered to students a powerful approach to understand the insights that arise from this area of enquiry. Too often, thermodynamics has moved away too far from macro chemistry to a world of representations and mathematical manipulations, leading to a confused generation of learners. He rooted his ideas on observing macro events.

He employed simple and accessible ideas. He sought to develop an understanding of the way energy was so central in understanding the behaviour of matter at the molecular level. In particular, the way he developed the ideas that lead to the concept of entropy allowed the learners to see the significance of the second law as the law that summarises the direction of change: a universal principle in every way as C.P. Snow (1905–1980) had argued so cogently long ago (Snow, 1956, 1959). Snow was able to see the much bigger picture being not only a physical chemist but also a successful novelist.

Johnstone's research showed that organic chemistry related to hydrocarbons was accessible to very young learners. He was able to relate this to the oil industry, the world of products deriving from hydrocarbons and then on to foods. This invoked little more than the idea that carbon often links to hydrogen, oxygen and nitrogen. Much of this was macro chemistry but it can be interpreted and understood using very simple representations. Functionality was left until later. It is possible to argue that organic chemistry needs further re-thinking to highlight the importance of functionality in determining organic reactions.

In all these areas, many new models were developed. Some were very simple. Some were physical, some were diagrams. The goal throughout was to reduce the demands on limited working memory resources so that genuine understanding of the molecular world might be gained.

The next chapter seeks to summarise the central meaning of Johnstone's triangle and the implications of the model for practical teaching.

## References

Duncan I. M. and Johnstone A. H., (1973), The Mole Concept in Chemistry, *Educ. Chem.*, **10**, 213.

Egan K., (1972), Structural communication - A new contribution to pedagogy, *Program. Learn. Educ. Technol.*, **1**, 63–78.

Garforth F. M., Johnstone A. H. and Lazonby J., (1976a), Ionic Equations and Examinations at 16+, *Educ. Chem.*, **13**, 41.

Garforth F. M., Johnstone A. H. and Lazonby J., (1976b), Ionic Equations - Difficulties in Understanding and Use, *Educ. Chem.*, **13**, 72.

Hassan A. K., Hill R. A. and Reid N., (2004), Ideas Underpinning Success in an Introductory Course in Organic Chemistry, *Univer. Chem. Educ.*, **8**, 40–51.

Johnstone A. H., (1974), Evaluation of Chemistry Syllabuses in Scotland, *Stud. Sci. Educ.*, **1**, 21–50.

Johnstone A. H., (2000), Teaching of Chemistry – Logical or Psychological? *Chem. Educ.: Res. Pract. Eur.*, **1**(1), 9–15.

Johnstone A. H. and Kellett N. C., (1980), Learning Difficulties in School Science - Towards a Working Hypothesis, *Eur. J. Sci. Educ.*, **2**(2), 175–181.

Johnstone A. H. and Morrison T. I., (1964), *Chemistry Takes Shape*, London: Heinemann Educational, vol. 1.

Johnstone A. H. and Morrison T. I., (1969), *Chemistry Takes Shape*, London: Heinemann Educational, vol. 5.

Johnstone A. H. and Webb G., (1977), *Energy Chaos and Chemical Change - A Bridge Book in Thermodynamics from School to University*, London: Heinemann.

Johnstone A. H., Morrison T. I. and Sharp D. W. A., (1971), Topic Difficulties in Chemistry, *Educ. Chem.*, **8**, 212.

Johnstone A. H., Morrison T. I. and Reid N., (1981), *Chemistry About Us*, London: Heinemann (reprinted 1982, 1986, 1988).

Johnstone A. H., Bahar M. and Hansell M. H., (2000), Structural communication grids: A valuable assessment and diagnostic tool for science teachers, *J. Biol. Educ.*, **34**(2), 87–89.

Miller G. A., (1956), The magical number seven, plus or minus two: some limits on our capacity for processing information, *Psychol. Rev.*, **63**, 81–87.

Oraif F. A., (2007), *An Exploration of Confidence Related to Formal Learning in Saudi Arabia*, PhD thesis, Glasgow: University of Glasgow, http://theses.gla.ac.uk/8033/, Accessed 9 November 2019.

*Chemistry for the Certificate of Sixth Year Studies, Memorandum 4*, Reid N. (ed.), (1982), Dundee: Scottish Curriculum Development Service.

Reid N., (2010), A Scientific Approach to the Teaching of Chemistry, *Chem. Action*, **90**, 29–44.

Research briefs, (undated), https://web.archive.org/web/20140221223911/http://gearup.ous.edu/sites/default/files/Research-Briefs/ResearchBriefSpiralCurriculum.pdf. Accessed, 16 March 2020.

Snow C. P., (1956), *The Two Cultures*, Cambridge: Cambridge University Press. [re-printed in 2010].

Snow C. P., (1959), *The Two Cultures and the Scientific Revolution*, Cambridge: Cambridge University Press.

CHAPTER 8

# *Bringing It Together: The Model and Chemistry Teaching and Learning*

## 8.1 Looking Back

In the first seven chapters, an attempt has been made to analyse the evidence that shows that chemistry is found difficult by many because its nature places high demands on limited cognitive resources. The working memory cannot cope easily.

Johnstone's triangle model brought this together in a simple but memorable way. This shows that, *from the perspective of the learner*, the nature of chemistry involves thought at three levels. Given the limitations imposed by working memory, it is essential that we do not invoke all three levels *at the same time*. An excellent way forward is to start with the macro and add on the other levels step-by-step. Johnstone developed numerous ways to do this.

The key for the teacher is always to remember that working memory capacity *controls* understanding. Bringing in the representational and the sub-micro must be carried out with care. Part of the key lies in the word *automaticity*. When the macro is so familiar that it can be seen as automatic, then there is little cognitive demand, leaving cognitive resources available to think at a second level.

In this chapter, an attempt will be made to synthesise and bring together the key findings to generate a coherent whole. Given that there are extensive references in previous chapters, this chapter will summarise with minimal referencing.

The summary will look at what Johnstone intended in developing the triangle model and what the model was never intended to be. The central

Advances in Chemistry Education Series No. 6
The Johnstone Triangle: The Key to Understanding Chemistry
By Norman Reid

Published by the Royal Society of Chemistry, www.rsc.org

goal in chemistry education is the development of understanding. Given the nature of chemistry as it explores matter at the molecular level, such understanding must involve mental models of reality. In developing understanding, the role of the working memory is critical.

The implications of the triangle model affect every aspect of chemistry teaching and learning from curriculum construction right across to assessment. The evidence in each aspect will be summarised.

## 8.2 The Research Story

With the development of updated school chemistry curricula in many countries from the 1960s, it became apparent that chemistry was seen as difficult by many learners. The new curricula had often attempted to move the emphasis away from memorisation (often: preparations and properties) towards an understanding of the world at a molecular level. It is sad feature of many textbooks today that there is frequently a subtitle that says something like '*all you need to know....*', the insidious pressures of examinations and the performativity agendas driving education *backwards* to the emphases of the mid-20th century.

We are in serious danger of undermining the value of education by the naive belief that examination outcomes are indicators of educational quality. Textbooks and teaching itself are now being driven by the need to maximise such performance and this is extremely damaging.

In his exploration of the areas where students found difficulties; Johnstone was able to demonstrate that the conceptual nature of chemistry now being taught in schools was placing high information demands on young learners. To understand the chemistry, the learners had to hold too many ideas in their working memories at the same time. Overload was the inevitable outcome and understanding became the casualty.

The '*expert*' had developed the skills to chunk several key ideas so that the working memory saw them as one, leaving capacity for further thought. The young learner lacked these skills and the working memory could not cope. Faced with examinations, the student was forced to memorise and attitudes to the study of chemistry started to deteriorate. Johnstone was able to demonstrate that the capacity of working memory *controlled* extent of understanding, a finding that is probably one of the most powerful educational insights of the 20th century. It affected all aspects of learning and assessment. His triangle model is a very concise summary of the implications of that insight in the context of chemistry. It brings together the nature of chemistry as a discipline and the way the human brain develop understanding of chemistry with the learners (Figure 8.1).

Johnstone taught chemistry for most of his career. This brought a reality to his research. He never lost sight of his central goal: how can we make chemistry more accessible to more school students and how we awaken a level of understanding with school and university students that allows them to see the remarkable human intellectual achievement as we have come to

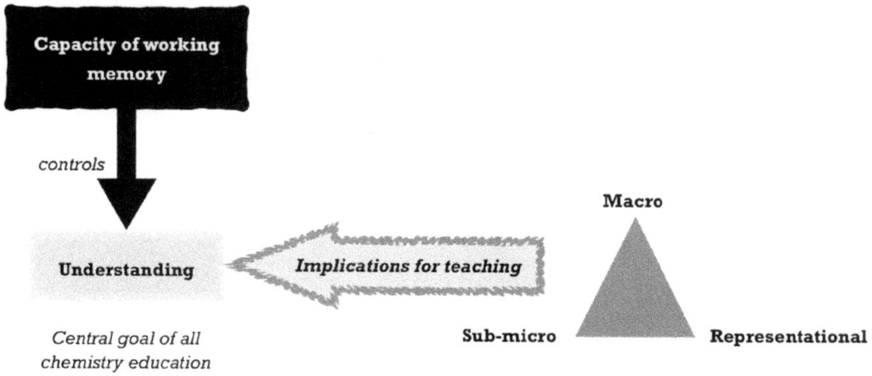


**Figure 8.1** The power of the triangle model.

understand the world at a molecular level, with all the implications that have arisen for the potential benefit of humankind.

His research was centred around the search for ways to make chemistry more accessible, asking why is understanding chemistry often demanding. This generated two sub-questions:

(1) Are difficulties caused by the nature of chemistry?
(2) Are difficulties caused by the way humans gain understanding?

Too often, we have naively argued that chemistry is difficult because many areas of the subject are abstract, and we have failed to recognise that abstraction is a *description*, not an explanation. Johnstone had now shown *why* abstract ideas were demanding for learners. Understanding a concept often require the learner to hold many ideas at the same time, to gain any kind of understanding. If the number of ideas to be held at the same time is too large, working memories cannot cope and understanding is simply impossible.

In thinking of the ways the nature of chemistry generated the difficulties for younger learners trying to understand the world at the molecular level, Johnstone realised that there were three levels of thought involved in seeking to gain that understanding (Figure 8.2).

His great insight was that he deduced that experienced chemists can move freely between the three levels, but the learner is unable to handle all three levels *at the same time.* This is simply a function of limited working memory capacity.

He was never seeking to develop some kind of philosophical analysis of chemistry as a subject discipline. From that perspective, maybe there are more levels. He was never arguing that the idea of three levels should be taught to those studying chemistry: indeed, that might frighten them off irreversibly! He was never arguing that our goal must be to develop links between the levels—that loses his central insight. He was never trying to

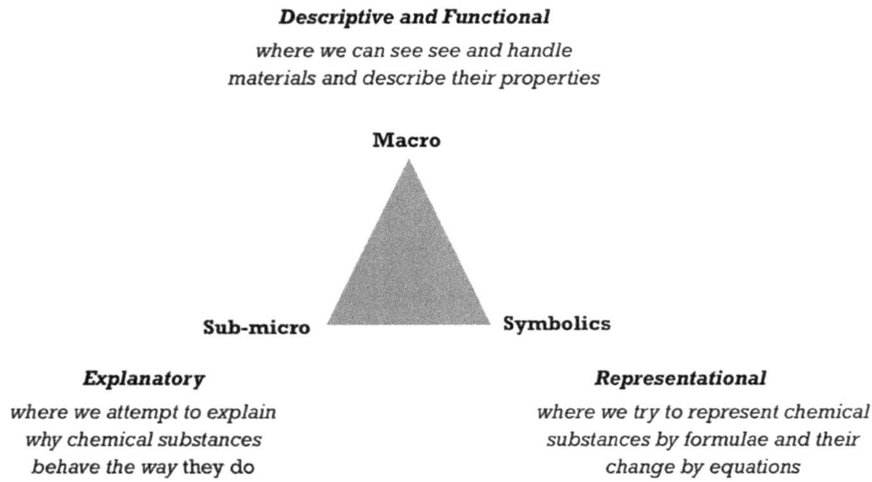


**Figure 8.2** Johnstone's triangle.

justify the use of models in the teaching of chemistry: mental models are central and physical models may or may not have a useful role to play in developing these.

His central argument can be seen as:

- Potential working memory overload is the *central* problem that undermines understanding.
- We should not try to work at all three levels *at the same time* with novice learners.
- At the outset, it is better to work at one level until ideas there are secure.
- Much excellent rigorous chemistry can be conducted at the macro (descriptive) level.
- Once the macro level is secure, we can bring in the representational and then bring in the sub-micro interpretation.
- The representational and the sub-micro can interrelate considerably in that the former can attempt to represent the latter.

Perhaps, the central finding of his research can be summarised as in Figure 8.3.

This central argument has immediate implications for the way school chemistry curricula are developed. It also has vital implications for the way chemistry is taught and, specifically, for the way textbooks are developed. In these areas, Johnstone has pioneered the way in his various textbooks that made the new chemistry curriculum in Scotland (first examined nationally in 1965) so successful. In addition, his quite remarkable thermodynamics book has shown that the same approach has direct relevance at university stages.

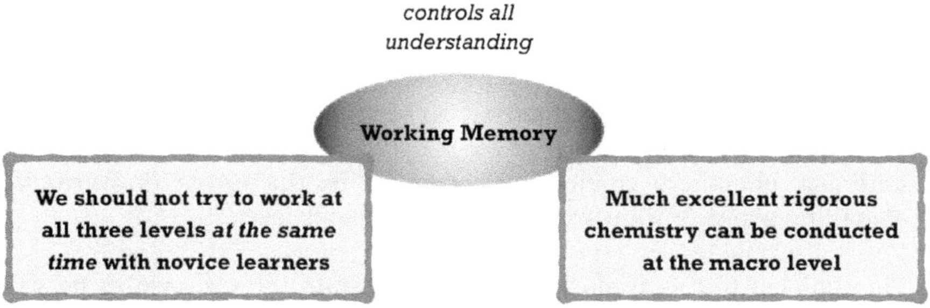


**Figure 8.3** The central argument.

Both laboratory work and group work have key places in the learning of chemistry. Laboratory work needs to be re-conceptualised as the place that can make chemistry real. In that sense, it can act as a bridge between the macro and the symbolic-representational levels. However, laboratory work has the potential to allow learners to gain insights into how chemistry gains its insights and the kinds of questions that can be asked of the macro world around us. All of this has major implications in the way laboratory work is employed at secondary stages. This might be an area where radical re-thinking is needed desperately. There have been useful developments at university levels but the purpose and emphasis of laboratory work at school level needs changed.

Johnstone always held the view that, at school level, chemistry had a vital part to play for all learners as part of their educational journey. Along with the other sciences, chemistry can help young people to make sense of the world around them. The findings of chemistry have made an enormous impact on lifestyles today and it is important that, as future citizens, every learner had some appreciation of this contribution. All our citizens need to be equipped to take informed decisions and some of the decisions have strong chemical features.

In much educational thought today, there is a repeated emphasis on *learner-centred learning*. Sadly, this idea is often left very confused. For some, it means learning built around learner-centred learning, usually with group work, problem-based learning, inquiry-based learning and discovery learning. The evidence shows that none of these approaches holds the key to better understanding. Indeed, while all the approaches hold considerable potential in other areas, there is often a failure to appreciate that the teacher is still central. The teacher has developed these learning situations. Indeed, the chemistry curriculum itself is usually not determined by the learners. It is developed by subject '*experts*'. Nonetheless, in the actual '*delivery*', learner-centred learning thinks more in terms of individualised or group work situations.

In a remarkable way, the triangle model actually does move the focus on to the learners. Therefore, in his model, Johnstone was focussing on the learner: why is chemistry demanding for most learners? How can we make chemistry accessible to all learners?

This leads on naturally to fresh ways of conceptualising the chemistry curriculum:

***Reject***: chemistry curricula determined by '*experts*' on the basis of the logic of chemistry.
***Embrace***: chemistry curricula determined by the needs of learners in making sense of their world at the molecular level.

This has led to the applications-led curriculum, now shown to be so effective as well as even more rigorous than traditional approaches.

We need to remember that the logic of a subject discipline like chemistry is artificial. It often does not reflect the tortuous story of the way chemistry revealed its insights historically. We are looking back at a more complete picture and imposing wisdom with hindsight.

The goal of school education is to equip future generations to develop their full potential, and to be able to be fully participating members of their society. The chemistry curriculum is there to offer a framework to bring maximum benefit to all learners. It is not there to force the learners into the often-artificial logic of a subject discipline as seen by experts in that discipline.

One major issue today is that of assessment. What happens in national (and international) assessment largely determines what is taught and what is learned. While curriculum development may have moved considerably over the decades, assessment has often not kept pace. The problem today is that assessment is being used today in ways that undermine its value.

National and international assessments are being used to '*grade*' schools and countries. This forces a trend to the reward of recall, and this makes the life of the imaginative school chemistry teacher intolerable. It is worth looking at the way countries that have adopted this line are also finding that teacher retention rates also decline. Even here, Johnstone's triangle has something to say: re-focus everything on the learner and the benefits that are possible from an education in and through chemistry.

## 8.3 The Triangle and Chemistry Education Today

In most countries, the chemistry curriculum and the way it is assessed in national examinations is determined mainly by those who are not practising teachers. Sometimes this makes the task of teachers very difficult in that they may be mandated to do what they know is probably not in the best interests of the learners. However, even within such restrictions, there is much that is open to the chemistry teacher.

In an ideal world, the chemistry curriculum would be re-structured so that the macro dominated at early stages and the curriculum design would allow for the systematic and step-by-step introduction of the representational and sub-micro interpretation. The assessment systems would be coherent with this approach and place emphasis on understanding ideas rather than recalling information. For this to happen, there would need to be considerable changes in the way curricula are developed and, even more importantly, in the way assessment is employed.

In the real world, however, the triangle model offers a powerful summary of the nature of chemistry from the learner perspective and emphasises that understanding chemistry is never going to be an easy task. The essential problem rests in trying to make sense of the world at the molecular level when none of the entities can be seen and, indeed, are not open directly to human senses. Experimental evidence (which is open to the senses) has to be interpreted and represented.

To achieve that we employ language. However, we also employ symbolisms, visual representations and models. Sometimes, we invoke analogy, with all the inherent dangers. All have their place. However, what the triangle model stresses is that the known limitations of the working memory capacities of the learners control the extent of understanding. Our teaching approaches must take this into account.

Johnstone's central argument was that the novice learner could not operate at the three levels *at the same time*. Therefore, we must start with the tangible, the macro, what can be experienced with the senses. Once the macro is well-established with the learners, then we can move step-by-step to introduce the sub-micro to interpret and explain what is observed. Inevitably, we need to represent the world at the molecular level in some way.

The central point is that, no matter what approach we adopt to introduce the sub-micro and representational to interpret and explain what is observed, the critical limitations of working memory capacity must be considered. This may involve explanation, analogy, diagrams, pictures, models, *etc*. Whatever approach is employed, the key is to introduce it step-by-step.

There are two fundamental principles implicit in the triangle model:

(1) Whatever we do must be presented in a way that does not overload limited working memory capacity.
(2) The goal is *mental* models of physical reality where the models fit the evidence and make some sense of it: the goal is correct understanding.

In all this, we need to set one single central goal: we want to enable the learners to observe and describe the molecular work and then be equipped to interpret and represent that world. At early school stages, the macro must dominate but we want to allow all learners to be able to look at their world through the eyes of the chemist to make sense of it.

## 8.4 The Triangle and Teaching

We saw that, somewhat artificially, we could divide teaching into three broad areas:

- *Teaching as telling:* teacher-developed materials and presentation.
- *Teaching as sharing*: various forms of teacher-developed group activities.
- *Teaching as doing*: various forms of teacher-developed practical activities.

We saw how research shows that none of these is in any way superior in terms of developing student understanding. However, each approach can bring wider benefits in different areas. A look at these three approaches shows us that they are all '*teacher-developed*'. The role of the teacher is central. We do not allow our students simply to do what they want, when they want, in the way they want. The learning situations are developed by teachers.

Teaching as telling is far more than just the traditional image of a teacher standing in front of a class and talking while the class listens (or perhaps does not listen!). Sharing is more than the dialogue of *question and answer* after a didactic presentation while *doing* in chemistry centres on laboratory work, but the nature of that work often needs re-thought. In his textbook of 1981, Johnstone *et al.*, (1981, p xi–xii) speaks of the place of group work:

> *The discussion sections in the book are designed to allow pupils to have their say in small peer groups. The teacher should not be the central figure, but should act as a catalyst as he moves from group to group.... Our research shows quite clearly that, .... when pupils are invited, through structured discussion, to take part in the decision process, considerable impact is made.*

In all three areas the Johnstone triangle has clear implications. It does not matter whether we employ class lectures, carefully structured group work, or sophisticated laboratory experiences. If we overload the working memory by invoking all three levels *at the same time,* then understanding will be greatly hindered.

We saw that research shows that the strengths of the three approaches can be summarised:

| | |
|---|---|
| *Didactic* | Greatest strengths lie in communicating information and developing understanding, as well as provoking thought. |
| *Group work* | Greatest strengths lie in the development of attitudes and the development of numerous valuable generic skills. |
| *Laboratory work* | Greatest strengths lie in making the chemistry real and allowing the experimental to answer questions about the chemical world. |

It is important that we employ all three broad strategies for their strengths. It is even more important that the messages from the Johnstone triangle inform the detailed ways by which we develop the learning experiences.

Johnstone was adept in developing simple but practical ways to enable young learners to make more sense of chemistry. In this, he had two great strengths. First, he was sensitive to the learners and was able to pick up quickly when they were facing difficulties. Much arose from the fact that he continued to teach chemistry for most of his working career. He never lost empathy for learners. Second, he was acutely aware of the central reasons why chemistry is difficult: the working memory cannot handle multiple levels *at the same time*.

Language places demands on working memory resources, especially when common English words are employed with precise and specific meanings in chemistry. Many analogies and models generate mental models that are simply incorrect. Later *un-teaching* is extremely demanding cognitively. He argued strongly that such wrong models must never be employed.

The areas of major difficulties in chemistry are known and there is much research that offers practical ways to reduce the problems. The task of the teacher today is to build their teaching around such approaches, wherever possible. However, the fundamental principle is that we want to allow students to develop mental models that reflect the experimental evidence and offer useful ways to make sense of the world at the molecular level. This is far more than just creating clever ideas.

In the last chapter, we looked briefly at some of the main areas where research has shown there are problems. The Johnstone triangle offers a guide in developing ways to reduce these problems:

> *The mole*: The key lies in seeing it as a useful counting device and then develop ways to make the mole concept tangible. The concept of '*power*' has been shown to be a powerful chunking device.
>
> *Redox*: All kinds of practical ways have been found to reduce working memory overload but, again, it is important to link the ideas to the experimental evidence: the macro.
>
> *Thermodynamics*: At all costs, we must avoid this area ending up as a set of mathematical manipulations unrelated to the world around us. In his book, Johnstone showed that this was possible.
>
> *Organic*: Johnstone showed how excellent models can be generated simply using balloons and this allowed confidence to grow. However, the issue of functionality needs further thought in the light of the triangle model.

Chemistry possesses two very powerful tools to enable the macro to be placed centrally. The first is the laboratory. We need to re-think the way we use laboratory learning. It can be powerful in making chemistry real—here is macro chemistry (albeit in an artificial way) at its best. However, we must allow time for the macro to develop the thinking needed to gain mental models of what is happening. Laboratory work is also powerful in enabling learners to see how chemistry gains its insights, but we must allow the laboratory learning to function in this way. The laboratory is where we address questions of the macro world around us and see what answers it provides by the way it reacts.

The second very powerful tool is the Periodic Table but here again the Johnstone triangle has much to teach us. By developing Periodic Tables that looked at only one property at a time, the working memory can cope. The key

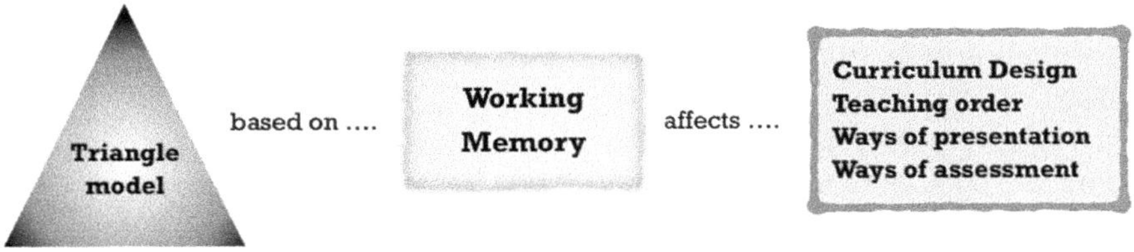


**Figure 8.4** Impact of the triangle model.

is then to relate that macro property to the sub-micro understanding, and this is where the Periodic Table can be immensely useful. Employing various representations (like graphs or pie charts), the patterns and trends can be made more accessible. The evidence shows that these approaches make sense to young learners and enable them to develop sound mental models.

The triangle model is based on the evidence related to the working memory. Its capacity is genetically determined and limited. The working memory is the only part of the brain where we think, understand and solve problems. Given the nature of chemistry as a subject discipline, its role is central in controlling understanding. Therefore, the triangle model affects every area of chemistry teaching and learning (Figure 8.4).

## 8.5 Ten Principles

Many years ago, Johnstone was asked to develop a new chemistry course for first year university students. He started by setting down a set of principles to underpin the new course. In fact, he listed ten principles. One of his students referred to these as the *'ten commandments'*—the nickname stuck (Johnstone, 1997). It is a sound way to approach curriculum development. Summaries, interpreted for school and university levels, can be found in Mbajiorgu and Reid (2006).

Here is an attempt to list the ten principles that derive from the Johnstone triangle. Most, but not all, are within some measure of teacher control. It is hoped that this set of principles will guide us forward so that the next generation of young people can enter into a satisfying and meaningful experience of understanding the world at the molecular level—one of the great achievements of human endeavour.

***Ten Principles***

(1) Working memory is of fixed and limited capacity and its capacity controls the extent of understanding: make awareness of this be your central starting point in all teaching.
(2) Learning chemistry can involve working at three thought levels: be aware that this places great burdens on limited working memory resources.
(3) Curriculum designers must be made aware of the model and re-think curriculum construction in line with human psychology.

(4) There is much excellent chemistry that can be taught rigorously at the macro level: start here.
(5) When the macro is well-established and ideas are *automated*, only then slowly introduce the other two levels.
(6) Change teaching order as necessary to avoid the introduction of the representational and sub-micro too early.
(7) Remember: the development of valid mental models of the behaviour of matter at the molecular level is the key goal.
(8) In linking the macro to the other levels, make sure that wrong models are not taught.
(9) Make sure that all assessment (school and national) reflects the central curriculum goals.
(10) As a discipline, chemistry is an example of great human endeavour with enormous scale implications: make sure that message is heard.

## References

Johnstone A. H., Morrison T. I. and Reid N., (1981), *Chemistry About Us*, London: Heinemann (reprinted 1982, 1986, 1988).

Johnstone A. H., (1997), Chemistry Teaching, Science or Alchemy? *J. Chem. Educ.*, **74**(3), 262–268.

Mbajiorgu N. and Reid N., (2006), *Factors Influencing Curriculum Development in Chemistry*, Hull: Higher Education Academy, pp. 18–19, https://www.advance-he.ac.uk/knowledge-hub/factors-influencing-curriculum-development-chemistry, Accessed 19 October 2019.

# Subject Index

www.ingramcontent.com/pod-product-compliance
Ingram Content Group UK Ltd.
Pitfield, Milton Keynes, MK11 3LW, UK
UKHW022101170926
8624IPUK00023B/19

9 781839 161681